T0401765

Circular Materials

Innovation and Reuse in Design and Architecture

gestalten

Contents

Reexamining the Value of Materials

What role can recovered and natural materials play in architecture and design?

There is an urgent need for a more circular future. The idea that we can throw away an object with scant consideration of its material footprint and environmental impact was contrived during the postwar economic boom of the mid-20th century. Mass production coupled with new and cheap plastics, which were pushed as abundant and disposable, significantly lowered the cost of consumer goods, making them more accessible than ever.

Plastic was initially marketed as a practical, resilient, and durable material. Industry soon realized, however, that more money can be made when it is discarded. Their messaging changed, asserting that it no longer made economic sense to invest time and effort in repairing and reusing goods given the growing cost of human labor and the low, "throwaway" price of plastic. This led to the rise of disposable goods such as cups, straws, and cutlery. Over time, the range of materials deemed by society acceptable to dispose of en masse has expanded to include textiles, leather, glass, ceramics, metals, e-waste, and more.

This book focuses on the innovative use of materials that are often discarded and perceived as being of lower quality or value compared with their conventional counterparts. The value of a material, however, is largely determined by our perception and understanding of it. For instance, Studio ThusThat (pp. 14 and 36) points out that in 2019, boron mines in California had been producing significant amounts of "waste rock" as a by-product of extraction—but on reevaluation, it was discovered that this supposed waste product contained substantial quantities of lithium, a crucial mineral for battery production. This revelation transformed the perception of the rock and enabled the site owners to extract considerable value from it.

Circular Materials is a compendium of case studies highlighting designers who have similarly reexamined materials and uncovered their potential value. The book aims to showcase designers around the world who work with these supposed "waste" streams and other regenerable materials, from design studios like Material Cultures, which has used locally sourced hempcrete in their Block House (p. 104), to manufacturers like Ananas Anam (p. 130)—which makes pineapple-leaf "leather" from vast amounts of agricultural waste in the Philippines. It investigates local designers working regionally with nonextractive, readily available materials, and international designers working at an industrial scale, throughout both low-and high-tech case studies that spotlight the use of reclaimed, recycled, and cultivated materials in architecture and design.

Current patterns of global consumption

Global waste generation still continues to rise. Unsustainable patterns of consumption and production have been powered by unchecked economic growth. As rapidly changing trends permeate all aspects of culture, the notion that our possessions should be replaced every few years, whether they are broken or not, is reinforced. Research by furniture designer CoCo Ree Lemery of Studio Kloak (p. 160) revealed that a "lifetime" warranty in the State of California may last as few as three years. Accordingly, when learning about the projects throughout this book we should remember that even "sustainably" manufactured furniture, clothing, and building materials are seldom truly sustainable if they are regularly disposed of in a linear system.

Planned obsolescence has further accelerated the disposal of goods, especially contributing to the ongoing escalation of e-waste. The world now generates about 2.3 billion tons (2.1 billion tonnes) of waste annually. Of this, 38% is disposed of in a completely uncontrolled manner—dumped in the environment or burned. An additional 30% of global waste ends up in landfill. This throwaway culture is fundamentally unsustainable and has led to significant threats to both environmental and human health. Scientific data shared in 2024 alone have exposed some of the most dangerous effects.

Above: A series of small-scale everyday objects from the Plastic Baroque series by James Shaw, made from recycled HDPE plastic. Opposite: Interior of the yoga studio by Invisible Studio, constructed with rammed stone.

> For the first time, the earth's average temperature over a 12-month period was more than 2.65 degrees Fahrenheit (1.5 degrees Celsius) above preindustrial times—exposing ever more people to catastrophic and unpredictable weather events.
>
> According to a study published in leading medical journal *The Lancet,* air pollution is now the leading contributor to the global disease burden, surpassing hypertension and smoking.
>
> Microplastics can be found everywhere, from the top of Mount Everest to the depths of the Mariana Trench, and even within human brain tissue.

The construction industry is a major contributor to our waste crisis. With rapid urbanization, continuous global development equates to the construction of a new city the size of Paris happening every week. In fact, the construction industry is responsible for over one-third of all solid waste and up to one-half of the world's material extraction, according to the World Green Building Council.

While waste and its impacts are hidden, out of sight and out of mind for many people living in the Western world, "end-of-line" countries such as Ghana suffer the consequences. Every day, 110 tons (100 tonnes) of unwanted clothing from wealthier countries are discarded at the Kantamanto market in Accra, one of the world's largest secondhand clothing markets.

Distant, lower-income countries supply many of the raw materials used in the West, and complex supply chains work to conceal the material origins of products from consumers. This makes it challenging for individuals to understand the environmental and social impacts associated with these resources and the realities faced by communities in these faraway regions.

The development and extraction of fossil-based materials disproportionately impact poor and marginalized communities with limited political power. The most heavily polluting and hazardous industrial facilities are typically concentrated in these areas, exemplified by places like "Cancer Alley"—an 85-mile (137-km) stretch along the Mississippi River. In that area, about 500,000 residents live in close

“The book investigates local designers working regionally with nonextractive, readily available materials, and international designers working at an industrial scale.”

“By rethinking how we use and value these materials, we can decrease waste levels, reduce environmental impact, and foster more resilient and just communities.”

proximity to approximately 150 petrochemical facilities, oil refineries, and plastics plants, which produce materials like PVC, described by Greenpeace as the most environmentally damaging plastic. Areas like this are known as "sacrifice zones," areas permanently impaired by environmental damage. They are associated with numerous human health issues, including higher rates of heart disease, respiratory illnesses, and cancer.

Throughout the book, projects like Red Mud by Studio ThusThat (p.14) expose the shrouded supply chains and muddy origins of materials like virgin aluminum. At the same time, projects like the Tane Garden House by Atelier Tsuyoshi Tane Architects (p.82) illustrate that a future based on more ethical supply chains is possible. Construction materials for the garden house are sourced within a 31-mile (50-km) radius, which helps to support local economies. It is vital that new supply chains are based on fair and equitable practices, which the projects in this book seek to promote.

Transitioning to a circular future

Friedrich Hayek, a progenitor of neoliberal economics, championed the idea that natural resources should be exhausted until collapse wherever this maximizes income, with the profits invested in new, emerging resources. Although this is a maximalist view that few thinkers would expressly voice, it is clear that our environmental crisis is directly connected to our habits of consumption and economic systems. A society centered on economic growth demands exorbitant levels of energy consumption, the extraction of natural resources, and increased greenhouse gas emissions.

Shifting away from this system will take some time and concerted effort. In the interim, we must find innovative solutions to manage and repurpose materials that have already been generated, while also incorporating natural materials that work in symbiosis

Opposite: The playful Circus Canteen by Multitude of Sins. Above: Piñatex × Cat Footwear, Eco-Intruder shoe in Moroccan Blue, made using discarded pineapple leaf fibers.

with people and the planet. By rethinking how we use and value these materials, we can decrease waste levels, reduce environmental impact, and foster more resilient and just communities.

There are, of course, challenges to working with some of the materials highlighted in this book. The readiness and availability of reclaimed and salvaged materials are uncertain. Matching supply and demand is not always straightforward, and our systems and supply chains need to change to facilitate this shift toward circular design.

This book offers a glimpse into how forward-thinking designers are navigating the transition toward a circular future. It features architectural studios like Lendager (p.240), which has mined local buildings in Copenhagen for secondhand construction materials, to companies like Orange Fibre (p.134), which has created circular fabrics from the by-products of the fruit juice industry. Although work clearly remains to be done, this book illustrates that a future economy rooted in sustainable practices and resource efficiency is essential for our planet, as well as feasible.

Recycled Materials

Recycling involves mechanical or chemical processes that transform materials at the end of their life into raw resources for new production, reducing our reliance on the extraction of finite resources.

In a circular economy, recycling plays a crucial role in minimizing the quantities of resources that make their way to landfill, as well as the energy required to manufacture new products. It is often seen as a last resort, however, because it consumes energy and can degrade material quality over time—unlike strategies of reuse, repair, or refurbishing, which seek to retain products at a higher-value state.

Opposite: This is Copper collection by Studio ThusThat. Above: The Wave Cycle stool by Harry Peck. Made from cheap recycled bodyboards, which often break and are discarded with only a single use.

Recycling has fallen short of expectations, becoming a feel-good solution. Yet the failsafe of recycling shouldn't be used to justify the production and consumption of more and more virgin resources; a significant amount of material currently in circulation can be effectively addressed through recycling efforts.

Some materials, like aluminum and glass, can be broken down and recycled almost infinitely without degrading, as part of a closed-loop system where there is little to no waste. In Europe, approximately 80% of aluminum is recycled each year, but growing demand means that a significant amount nevertheless must be produced from raw material.

The processes of mining, refining, as well as smelting aluminum are highly energy-intensive. Recycling preexisting aluminum, by comparison, requires 95% less energy.

Thinking in loops

While some materials can be recycled almost endlessly, others—such as paper and plastics—face limitations due to material degradation. In more widespread, open-loop recycling, the materials processed degrade over time to the point where they can only be downcycled into products of lesser quality and value, or rejected as waste. In this sense, an open-loop system can be viewed as either semicircular or close to linear, depending on the efficiency of the recycling process. For example, cardboard and paper can be recycled as much as five to seven times before the decline in quality triggers a tipping point. Most recycled plastics are mixed with some percentage of virgin content, but even then, they can typically be recycled only two or three times before the critical quality threshold is reached.

Nave 1, 2, and 3 by Shane Schneck, a series of vases made from 100% postconsumer aluminum.

The urgent need for more circular practices

Despite decades of awareness, the recycling rate for plastics remains shockingly low. Research published in 2017 found that only 9% of all plastic ever produced has been recycled. This is attributed mainly to the contamination caused by labels, food remnants, and other substances as well as the difficulty and costs involved in collecting and sorting different types of plastics. National Public Radio reports that petrochemical industries have for decades spent millions of dollars misleading consumers on the feasibility of recycling at scale. In September 2024, the State of California filed an unprecedented lawsuit against ExxonMobil for allegedly deceiving the public by attempting to convince consumers that recycling could solve the world's plastic-waste crisis.

The Plastic Baroque dining chair by James Shaw, made from recycled HDPE plastic.

Designers have found innovative solutions to the challenges of recycling, with case studies on projects that leverage technology, innovations in material science, and new methods of production, recovering waste that would otherwise be downcycled or discarded. Dutch Design studio The New Raw (p. 16)

"By transforming overlooked materials and agricultural by-products into valuable resources, designers demonstrate the immense potential of recycling as a tool for creativity and sustainability."

The Swedish Stockings Table Mini by Gustaf Westman is a handmade side table containing between 80 and 350 pairs of tights diverted from landfills.

Above: The Pots Plus collection by The New Raw explores the possibility of using plastic waste in 3D-printed furniture. Top: Recycled plastic pellets from the same project.

utilizes a robotic 3D-printing process to recycle local municipal plastic waste into benches and public street furniture. James Shaw has produced the plastic-extruding gun, which turns the industrial extrusion process, usually performed by enormous, expensive machines, into a localized, hand-driven action (p. 42).

By transforming overlooked materials like red mud, copper slag, and agricultural by-products into valuable resources, these designers demonstrate the immense potential of recycling as a tool for creativity and sustainability.

Working with stigmatized and hazardous materials

Studio ThusThat challenges the value of "waste" with their Red Mud tableware collection by recycling the industrial leftovers of aluminum production.

Project
Red Mud

Designer
Studio ThusThat

Material
Red mud

Resource Type
Industrial waste

Country of Origin
United Kingdom

Manufacturer
Studio ThusThat

Area of Use
Product design

Category
Recycled

Aluminum is everywhere, from airplane parts to kitchen foil. But its transformation from raw material to shiny metal is a highly energy-intensive process that leaves behind a toxic by-product known as red mud. An estimated 4.4 billion tons (4 billion tonnes) of red mud are stockpiled globally in enormous landfills that can be seen from space, and more than 165 million tons (150 million tonnes) are generated every year—causing serious environmental concern.

Aluminum ore is mined mainly in China, Brazil, Australia, Guinea, and Jamaica, and shipped to refineries around the world, dissociating the final product from its muddy origins.

Despite existing methods of treating red mud, only 3% is used due to its stigma as a toxic waste. Studio ThusThat was interested by the scale and hidden nature of this issue. Working alongside scientists, they aimed to draw attention to the impact of aluminum and its shrouded industrial processes by turning the red mud into safe-to-use ceramic tableware, thereby contrasting the ceramic production process, which is associated with "fragility" and "finesse," with the bulky, massive-scale production of aluminum.

Print your city! 3D-printing Rotterdam's plastic waste

Single-use plastic is used momentarily by consumers, but endures in landfill for hundreds of years. The New Raw reexamines this relationship by recycling municipal waste, ensuring that plastic stays in the use cycle for longer.

Project
Knotty and Pots Plus

Designer
The New Raw

Material
Plastic waste

Resource Type
Municipal waste

Country of Origin
Netherlands

Manufacturer
The New Raw

Area of Use
Furniture design

Category
Recycled

Sustainable design is often seen as a secondary option or an afterthought, but The New Raw believes it should be the default. By incorporating locally sourced, recycled, and recyclable materials, they help address the challenges of global waste while creating new opportunities for creativity.

The designers have collaborated with local recyclers and municipal organizations to source plastic waste for 3D printing. Through the Print Your City initiative, they have worked with communities to co-create public furniture like the Pot Plus series, connecting people directly to the recycling process and engaging the community in sustainability efforts.

Robotic 3D printing allows precise control in the form and strength of each furniture piece while embracing the raw, recycled nature of the material. "The inherent textures and imperfections of recycled plastic," The New Raw declares, "can be turned into a design asset." As an example they cite the weave-like surface of the Knotty bench, which, like any hand-knitted piece, features slight differences due to the material's properties and variables in the production process.

MURAKAMI
THE WIND-UP
BIRD CHRONICLE

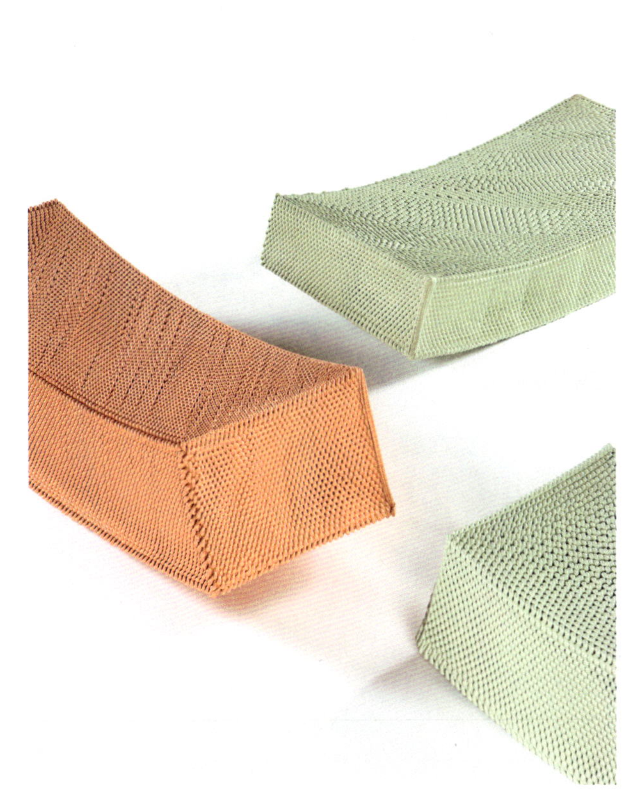

Knotty is a playful bench collection with bold textures inspired by knitting techniques. Thick knots create a tactile, permeable surface for water drainage without the need for internal structures.

Conserving limited resources and working within planetary limits

Each year, as many as 8 billion pairs of tights and pantyhose are purchased, worn only a few times, and quickly discarded—many of them made from petrochemical-based fibers such as nylon.

Project
Tights to Tables

Designer
Gustaf Westman

Material
Recycled tights and fiberglass

Resource Type
Textile waste

Country of Origin
Sweden

Manufacturer
Swedish Stockings

Area of Use
Furniture design

Category
Recycled

To address the growing volume of pantyhose and tights ending up in landfills, Swedish Stockings launched a recycling program, collecting old nylon hosiery of any brand from customers by mail.

Initially, the company downcycled the collected stockings, shredding them to make filler material for industrial fiberglass tanks. Ultimately, however, the company opted to team up with Swedish designer Gustaf Westman to create a collection of five different made-to-order tables.

Each table has a unique pattern and contains between 80 and 350 pairs of recycled nylon tights and pantyhose. The manufacturing process involves shredding the stockings and mixing them with recycled fiberglass. This material is then pressed into cylindrical molds, polished, and sanded by hand to achieve a smooth, marble-like finish. The collection aims to extend the lifespan of stockings by creating pieces that are valued for decades.

Miró
MÅNGA
BJÖRN WESSMAN
Norstedts

The Tights to Tables collection aims to extend the lifespan of stockings by creating pieces that are valued for decades.

Shifting perspectives on repair and restoration

Sculpted in Max Lamb's workshop, Box is a collection of 33 different furniture pieces made from leftover cardboard and a homemade wheat paste.

Project
Box

Designer
Max Lamb

Material
Cardboard

Resource Type
Commercial waste

Country of Origin
United Kingdom

Manufacturer
Max Lamb

Area of Use
Furniture design

Category
Reduced

Combining cardboard with a plant-based glue of flour and water, Lamb reconstructs the humble box. His low-tech collection is plastic-free, endlessly recyclable, and can be easily repaired using readily available materials.

The collection is crafted entirely from the design studio's own leftovers, using stockpiled cardboard packaging, toilet rolls, and miscellaneous boxes. Some boxes are left whole and reinforced internally with ridged cardboard lattices for additional strength. Offcuts are transformed into pulp using the wheat-based glue, applied in as many as 15 layers to create a supportive outer shell.

Lamb's approach to durability is one of pragmatism—the pieces are designed to be long-lasting but if the furniture does start to show signs of wear, he encourages the owner to patch it up themselves. You will simply need additional cardboard and some flour from the kitchen. Of course, there will always be concerns around the permanence of these objects, but Lamb argues that we need to learn that "things don't have to be perfect forever."

MAX

Occasional and intuitive stripes of paint—made from mineral pigments and linseed oil, rather than petroleum—offer the furniture's sole decorative touch.

The
IVO

Seven designers celebrating the versatility of aluminum

Hydro Circal 100R is the first industrial-scale aluminum material made entirely of post-consumer scrap, with a carbon footprint 97% lower than the global average for aluminum production.

Project
Circal 100R

Designers
Andreas Engesvik, Rachel Griffin, Max Lamb, Philippe Malouin, Shane Schneck, Inga Sempé, and John Tree

Material
Aluminum

Resource Type
Industrial waste

Country of Origin
Norway

Manufacturer
Hydro

Area of Use
Product design

Category
Recycled

Aluminum is well suited to the circular economy as it can be recycled infinitely without any loss in quality. To produce Circal 100R, aluminum sourced mainly from old buildings and used cars is shredded before rounds of sorting to remove unwanted alloys. The sorted chips are fed into a delacquering machine to remove any surface treatment before being melted and cast into billets, which can be used to make new products. Because the aluminum is sourced from post-consumer scrap, its carbon footprint is much smaller; its emissions have already been accounted for within the material's first life cycle. Pre-consumer scrap, by contrast, reuses only waste from the manufacturing processes, waste which has yet to complete a full life cycle.

For Milan Design Week 2024, Hydro tasked a group of seven designers to produce a range of furniture pieces using only extruded Circal 100R sections. To showcase the functionality, versatility, and minimal industrial aesthetic of aluminum, the manufacturer selected designers with different backgrounds and sensibilities. The monolithic pieces each have different personalities, but all feature interlocking profiles that slot together and eliminate the need for additional mechanical fixings.

"To showcase the functionality, versatility, and minimal industrial aesthetic of aluminum, the manufacturer selected designers with different backgrounds and sensibilities."

Color is later applied to the products through an anodizing process, which ensures the material can be endlessly recycled without any loss in quality. At the exhibition, John Tree presented a minimalist chair, testing the limits and complexity of aluminum extrusion using only two 8-inch (200-mm) dies. The legs and back support, which are bent in different directions, are made from one profile, connected but partially separated. Another profile is mirrored to form the seat and backrest. The exhibition also featured the Prøve light by Max Lamb, Grotte table lamp by Inga Sempé, Tsuba coat stand by Andreas Engesvik, T-Board shelving system by Philippe Malouin, and the Nave vessels by Shane Schneck, Serial modular room partition by Rachel Griffin.

Project
Structural Skin

Designer
Jorge Penadés

Material
Leather

Resource Type
Textile waste/bio-based

Country of Origin
Spain

Manufacturer
Jorge Penadés

Area of Use
Product design

Category
Recycled

A composite material made from discarded leather scraps

On recognizing the extent of waste generated by the fashion, automobile, and furniture industries, Jorge Penadés sought to rethink how we recycle leather, without using harmful resins.

Penadés developed a method for reconstituting discarded hide into a composite material, named Structural Skin, that expresses the unexpected beauty of leather offcuts.

To produce this material, shredded scraps of leather are combined with a bone glue—a biodegradable binder which, like leather, is a by-product of the meat industry. This mixture is compressed into iron molds and left to set before being impregnated with shellac, a natural resin derived from insects. The result is a solid material that can be worked like wood, and when sanded down reveals vivid, marble-like patterns created by the leather strips.

To ensure the biodegradable elements could be recovered at the end of life, Penadés initially used brass brackets, along with nuts and bolts to hold components together. He has since explored new systems with non-fixed connections, using ratchets, belts, and straps to join the boards of Structural Skin with blocks of glass, marble, and aluminum.

Translating material science out of the lab

Studio ThusThat has collaborated with material scientists at KU Leuven university to explore the various ways in which slag, the by-product of copper production, can serve as a low-carbon alternative to cement.

Project
This Is Copper

Designer
Studio ThusThat

Material
Slag

Resource Type
Industrial waste

Country of Origin
Belgium

Manufacturer
Studio ThusThat

Area of Use
Furniture and product design

Category
Recycled

Copper is crucial for an electric future: one wind turbine contains around 5.5 tons (5 tonnes) of copper, and 11 tons (10 tonnes) are used for every kilometer of high-speed railway. However, grades of ore are rapidly diminishing—ore mined in the 20th century was of about 40% copper purity, compared to about 0.2% purity now—and there is a real risk that accessible copper could soon be entirely depleted.

The demand for copper comes with an increased impact for the environment and the exploitation of low- and middle-income countries. Although the material's origins are often concealed, the metal we know as copper is actually part of a much wider material ecosystem that leaves behind a number of by-products. This Is Copper is a series of investigations into the use of slag, a by-product formed from the molten impurities cast aside during the smelting process, and produced during both the extraction and recycling of copper.

The collection contains lamps, mirrors, and the Sparkly Black chair, which explores the use of slag as the binder, filler, and formwork for a furniture piece.

Project
Potentials

Designer
Studio Johanna Seelemann

Material
Dead-stock denim

Resource Type
Textile waste

Country of Origin
Netherlands

Manufacturer
Studio Johanna Seelemann

Area of Use
Furniture design

Category
Reused

Shaping new pathways for end-of-life recovery

As part of the Art of Raw platform, Potentials explores the tactile qualities and structural stability of dead-stock denim. Each piece is inspired by the architecture and furniture of G-Star's Dutch headquarters, designed by OMA and Jean Prouvé, respectively.

The Potentials collection includes five pieces of office furniture that reinterpret conventional typologies and engage with circular material life cycles. Raw, untreated denim is supported by an indigo-colored wooden frame made of ash—interfacing in a simple, reversible manner that enhances the material's future reuse potential. The objects present different ways in which the fabric roll can be used and incorporated into the structure, through various nonpermanent methods of wrapping, stapling, and filling.

The stool, leaning bench, and room divider examine the principle of borrowing—making minimal alterations, cuts, and seams to the denim to maximize the viability of reuse. For the room divider, Seelemann developed a novel technique to fold and clip the denim roll into the timber structure, conserving material and enabling end-of-life recovery. The floor mat reimagines a typical anti-fatigue mat, designed to improve worker comfort when standing for long periods. Finally, the valet stand is a two-part structure used for hanging clothing items and explores an interplay of weight.

Drawing from the nomadic roots of the Kazakh region

Embracing locally sourced materials, the 10 Nodes Armchair is made using only three widely available components: the wooden handles of common garden tools, recycled plastic bottles, and a woollen felt.

Project
10 Nodes Armchair

Designer
Daniyar Uderbekov

Material
Wool, recycled plastic bottles, and shovel handles

Resource Type
Mixed

Country of Origin
Kazakhstan

Manufacturer
Daniyar Uderbekov

Area of Use
Furniture design

Category
Reduced

Daniyar Uderbekov advocates for the use of local materials and supply chains, calling for the dismantling of heavily polluting, industrial-scale furniture production. His innovative armchair is predicated on the idea of localized assembly, inspired by his cultural heritage and the construction of nomadic Kazakh yurts.

The chair's frame is made from common shovel handles with a diameter of 1.6 inches (4 cm)—chosen for their durability and accessibility across hardware stores worldwide. The handles are joined by 3D-printed nodes made from recycled single-use plastic bottles. Named for its resource-efficient design, the chair's frame relies on just 10 connection points in total.

The frame is covered with a woollen felt, a regenerative, biodegradable material that is ethically sourced and robust. While hand-rolling the felt, Uderbekov observed traces of grass, seeds, and threads embedded in the fibers, inspiring him to add vibrant specs of color to the chair's surface. The result is a rich, terrazzo-like pattern that celebrates the natural beauty of the wool.

Embracing the tactile qualities of recycled plastic

Typically, plastics are discarded after a single use, and barely given a second thought. James Shaw aims to change our relationship with plastic by exploring its materiality using an extruding gun.

Project
Plastic Baroque

Designer
James Shaw

Material
HDPE plastic

Resource Type
Industrial waste

Country of Origin
United Kingdom

Manufacturer
James Shaw

Area of Use
Furniture and product design

Category
Recycled

Shaw's playful furniture is crafted from the leftovers of plastic recycling plants, specifically high-density polyethylene (HDPE). This material often comes in the form of floor sweepings, or contaminated batches unfit for further use. In the United Kingdom, semi-skim-milk bottle caps are green, and whole-milk caps are blue, both HDPE. If these colors are mixed during the recycling process, however, the batches are often discarded. Through this outdated system, which should by now be obsolete, tons of material are made available to Shaw.

Some argue we should not use plastic at all, but Shaw sees the issue as more nuanced. The reality is that we have an overwhelming amount of plastic to work with—in 2021, the total amount of plastic waste produced in the European Union was more than 17.6 million tons (16 million tonnes), only 7.23 tons (6.56 million tonnes) of which were recycled.

While Studying at the Royal College of Art in London, Shaw created three handheld "guns," or tools, to produce objects, driven by the idea that new tools could unlock new outcomes and possibilities for untapped resources such as the plastic bottle caps. His best-known tool, the plastic-extruding gun, turns the industrial extrusion process into a localized, manual action.

"James Shaw draws inspiration from the Baroque movement, which celebrated earthly abundance, finding parallels with our current use of plastic."

Small plastic pellets are fed into the gun, heated, and extruded as molten plastic about 1 inch (2.5 cm) in diameter. Shaw's machine celebrates the qualities and materiality of plastic, aiming to change the way we think about it as a material. We might admire the grain on a piece of timber or the veins in a slab of marble, but rarely do we look at a milk bottle cap or shampoo bottle and think, "Now that's a lovely piece of polyethylene!"

The Plastic Baroque series is produced with the extruding gun and includes tables, chairs, and candlesticks—the last a by-product of the larger pieces. Shaw draws inspiration from the Baroque movement, which celebrated earthly abundance, finding parallels with our current use of plastic. Instead of viewing plastic as just a disposable material, we should recognize its richness, he argues. By appreciating what is around us, he suggests, perhaps we can use plastic more sustainably.

Project
Wave Cycle

Designer
Harry Peck

Material
Bodyboards

Resource Type
Recreational waste

Country of Origin
United Kingdom

Manufacturer
Harry Peck

Area of Use
Furniture design

Category
Recycled

Extending material lifespans with circular furniture

Environmental charity Keep Britain Tidy estimates that more than 16,000 cheap, low-quality polystyrene bodyboards are abandoned on UK beaches each year.

Manufactured in China and shipped thousands of miles to beach areas, like Devon in southwest England, these low-quality boards often break within a single use. The bodyboards are composed primarily of expanded polystyrene with a nylon coating; once the thin coating splits, the polystyrene breaks into tiny balls that pollute the oceans, take 500 years to decompose, and are commonly mistaken for food by marine life.

Polystyrene is a thermoplastic, which means that it can be melted and remolded, lending itself to recycling. Harry Peck performed more than 100 tests to develop a clear understanding of how best to process the material, eventually breaking the furniture production into three stages. The polystyrene is first melted down with heat, then shredded into small granules, and finally extruded into metal molds while heat is applied.

This circular process allows Peck to extend the lifespan of a typically short-lived material, transforming it into durable furniture pieces and keeping it within the manufacturing cycle. The flatpack collection consists of a stool, a chair, and a shelving unit, all made from recycled bodyboards.

Addressing complex relationships between humans and materials

We+ aims to reframe our relationship with waste through their Refoam collection, made with recycled waste polystyrene from Tokyo, where large quantities are discarded at fish markets and food stores.

Project
Refoam

Designer
We+

Material
Polystyrene

Resource Type
Plastic waste

Country of Origin
Japan

Manufacturer
We+

Area of Use
Furniture design

Category
Recycled

Typically, the material would be melted down into small blocks known as ingots and distributed globally to manufacturing sites—from which, in some cases, it is transported back to Japan in the form of plastic products. Though recycling rates were high, design studio We+ set out to reconsider these overcomplex logistical relationships. As an alternative, they explored using intermediate treatment plants to manufacture furniture from recycled polystyrene locally, in Japan—eliminating the need for long-haul transportation.

The collection features chairs, tables, and benches with a molten-like texture that expresses the recycling process of melting down and compressing the softened material into molds. Due to the polystyrene's thermoplastic properties, as the foam expands and cools within the tightly packed metal molds it becomes rigid. The undulating lines and bubbling texture of the polystyrene contrasts with the crisp, well-defined edges of the overall form.

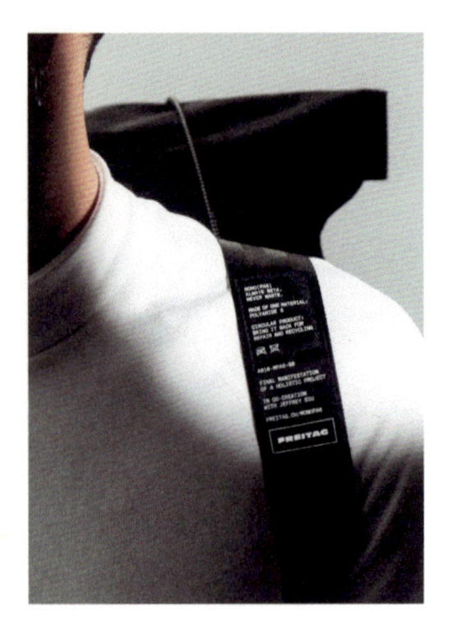

Project
Mono [PA6] Backpack

Designer
FREITAG and Jeffrey Siu

Material
Polyamide 6

Resource Type
Plastics

Country of Origin
Vietnam

Manufacturer
FREITAG

Area of Use
Fashion

Category
Reduced

Facilitating ease of recycling through mono-material solutions

To enable a simplified process of recycling that avoids a need for complex dismantling, the FREITAG Mono [PA6] bags are made from a single material: polyamide 6, better known as nylon.

Each of the bag's 17 components—from the zippers to the carrying straps, labels, and sewing thread—has specific performance requirements and properties, giving rise to various challenges. The bag's outer fabric, for example, needs to be water-repellent. Such waterproofing had conventionally required an additional coating or membrane made from other materials, but FREITAG and a partner company developed an innovative, water-repellent, three-layer textile made entirely from PA6.

The bags contain other trace materials, such as printer ink for the inside label and spots of adhesive, which are not made of PA6. Glass fibers are used in the zipper slide for additional strength. The proportion of these additional materials are so negligible, however, that the bags can be mechanically recycled into new PA6 once they can no longer be worn or repaired. FREITAG states that recycling tests have been completed at a material technology facility in Zurich, confirming that the bags can be shredded and then extruded into new PA6 granules, usable to produce components for new items.

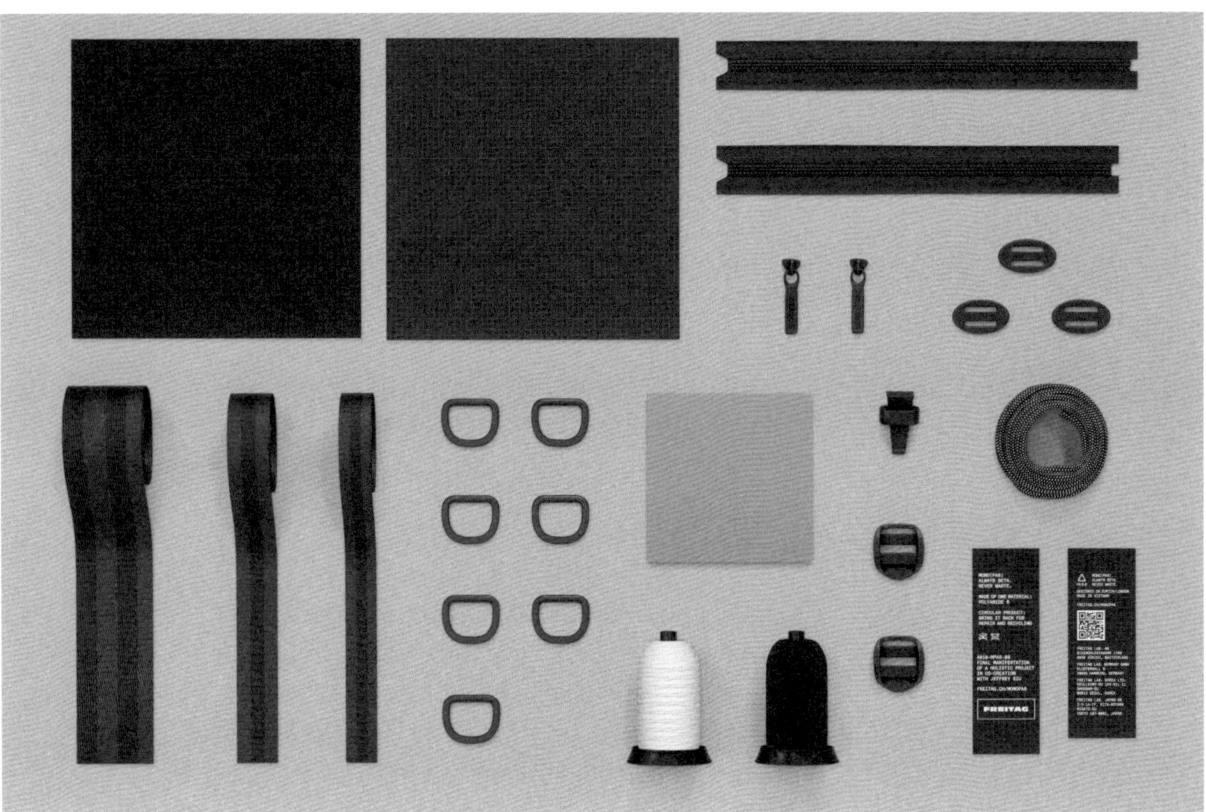

FREITAG has been thinking in cycles since its inception in 1993, when it first made messenger bags from used truck tarps. The company now aims to further slow and close resource loops and are making strides towards a circular economy.

Blowing a renewed life into cheap PVC pipework

Designer Kodai Iwamoto uses the heritage craft of glassblowing to transform PVC pipework, often used for plumbing, into unique handmade vases.

Project
PVC Handblowing Project

Designer
Kodai Iwamoto

Material
PVC pipework

Resource Type
Construction waste

Country of Origin
Japan

Manufacturer
Kodai Iwamoto

Area of Use
Product design

Category
Reduced

After a museum visit, Kodai Iwamoto considered how many of its ancient artifacts, though now revered, were once just everyday items. He became interested in our perception of value and how we can enhance the value of everyday objects. Naturally, he turned his attention to PVC pipework, which he had long used to make prototype models—but now he recognized its potential as a work of art in its own right.

Leveraging the thermoplastic properties of PVC, Iwamoto combines the cheap, mass-produced material with techniques inspired by glassblowing. He uses an infrared heater on the PVC pipe until it becomes soft and malleable. The pipe is then placed in a wooden mold and air is pumped through it, causing the plastic to expand and take the shape of the mold.

This process results in unique vessels, with their form influenced by various factors, including the mold design, the degree of air pressure, and how quickly and evenly the PVC is heated. By upcycling pipes from construction sites, Iwamoto hopes to reduce the amount of waste PVC going to landfill.

Project
Crystal Shell Collection

Designers
Shakúff

Material
Recycled Pyrex glass

Resource Type
Industrial waste

Country of Origin
United States

Manufacturer
Shakúff

Area of Use
Lighting design

Category
Recycled

A closed-loop recycling process recovering leftover Pyrex glass

The Crystal Shell Collection honors natural forms with its glinting, mollusc-like pendants. Joseph Sidof, the founder of Shakúff, spent childhood days on the shore, inspired by the natural world—whose motifs continue to suffuse his work.

Shakúff works with leftover Pyrex borosilicate glass cullet, melting crushed glass and hand blowing it to suggest organic shapes. A closed-loop recycling process imprints a speckled finish across the luminaires' surfaces and imparts an ephemeral quality to their edges. Illuminated with LED, the pairs of shell-shaped discs are fastened only by a brass fitting at the top, allowing easy access to components such as the driver. This allows electrical fittings to be maintained or replaced, rather than having to discard the entire lighting fixture.

Commercial glasses are typically divided into two types: soda-lime-silica and specialty glasses; the majority of glass goods produced are made from the former, including common drinking glasses and jars. Some speciality glass, including Pyrex borosilicate, is not typically considered recyclable, due to a higher melting point than soda-lime-silica glass, which prevents it from being melted as part of a larger, mixed batch, requiring instead a more expensive, small-batch process.

Project
aquaplus re

Designers
Lamy and HolyPoly

Material
Recycled plastic

Resource Type
Electronic waste

Country of Origin
Germany

Manufacturers
Lamy, HKS, and Varia Color

Area of Use
Product design

Category
Recycled

Embracing the material revolution, becoming a catalyst for change

The writing instrument manufacturer Lamy has collaborated with professional service firm HolyPoly to transform its watercolor-paint boxes into a more sustainable product using recycled plastic content.

Lamy aims to replace all virgin plastics with recycled material, across their entire product range. The aquaplus re, a watercolor-paint box, is their first product to phase out primary-source petroleum plastic. Together with HolyPoly, Lamy made a product composed of 99.4% recycled materials. The only nonrecycled content in the paint box comes from the 0.6% that consists of color pigments, which are not derived from plastic. The production process utilized the existing stock of machinery, with injection-molding tools.

The plastic used is a polystyrene that is sourced from discarded refrigerators. This switch has not only reduced the proportion of environmentally harmful greenhouse gas emissions by 69%, but also helps to address the issue of electronic waste—one of Europe's fastest-growing waste streams. Annually, Lamy also sells up to nine million writing instruments, about seven million of which are made from plastic. By switching to recycled materials, they aim to drastically cut carbon emissions, without increasing prices.

LAMY
aquaplus
12
12
aquaplus
Malen und Zeichnen mit Lamy

Rethinking outdated production methods in the wool industry

An innovative robotic system capable of manufacturing three-dimensional structures with wool industrially for the first time, without the use of water or additional materials in the felting process.

Project
Flocks Wobot

Designer
Christien Meindertsma

Material
Wool

Resource Type
Bio-based

Country of Origin
Netherlands

Manufacturers
TFT and Christien Meindertsma

Area of Use
Various

Category
Cultivated

Christien Meindertsma worked with robotics company TFT to develop the Flocks Wobot—a machine that operates like a 3D printer, but instead of layering filament, uses a needling system to felt wool into three-dimensional forms.

When Meindertsma was previously commissioned to assess what to do with 6.6 tons (6 tonnes) of discarded Dutch wool, she discovered that shepherds throughout the Netherlands were throwing wool away due to its perceived poor quality. European varieties of wool are often dismissed as too coarse for industries like upholstery, in favor of softer wool from New Zealand. Meindertsma found, however, that the problem is not the wool itself, but rather comes from limited and outdated production methods.

The Wobot seeks to repurpose the 1,650 tons (1,500 tonnes) of wool discarded each year in the Netherlands, at the same time offering a more sustainable alternative to synthetic materials like foam rubber, glass wool, and polystyrene foam. Meindertsma envisions the soft yet strong woollen forms having a spectrum of uses, such furniture, acoustic products, and insulation.

Project
Restless Textiles

Designer
Lobke Beckfeld

Material
Cotton fibers

Resource Type
Textile waste

Country of Origin
Germany

Manufacturer
Lobke Beckfeld

Area of Use
Fashion

Category
Recycled

A forensic material study exploring undervalued fibers

Working with textile waste from Frottana, a German towel manufacturer, Lobke Beckfeld proposes an experimental system for reprocessing three types of textile by-product: long-yarn waste, selvedges, and short-fiber remnants.

Beckfeld describes how "the evaluation of matter as 'waste' is based on subjective and societal perceptions and is not an intrinsic property of the material." Her experimental project Restless Textiles explores the future use of pre-consumer industrial textile waste. Through a series of prototypes, the project presents radical strategies for the recycling of natural fibers, producing valuable resources as an alternative to conventional downcycling.

Restless Textiles establishes a rules-based system for managing three categories of cotton waste, aiming to triple the fibers' useful life. First, warp rests—the long-yarn waste left after weaving—are sorted, bundled, and weaved into new textiles using a handloom, a meticulous process that requires significant time and effort, limiting its scalability. Second, yarn waste from selvedges is transformed into a nonwoven felt material. Finally, small-fiber waste up to 2 inches (5 cm) in length—produced during yarn preparation and too short to spin into new threads—is mixed with natural binders such as pectin to create a composite material. This composite serves as a protective coating for the nonwoven felt, further enhancing its utility.

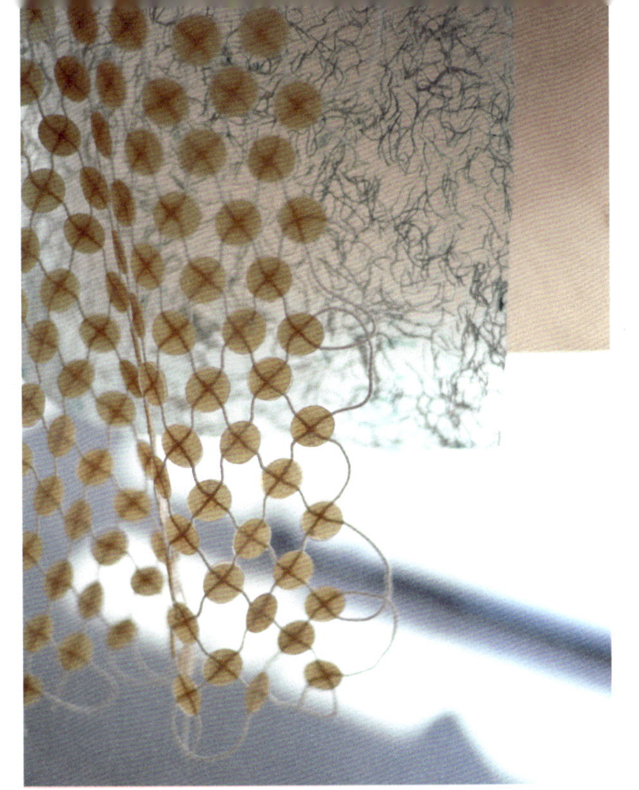

The materials are envisioned for use in decorative interior elements as well as in fashion—particularly seasonal or runway fashion, allowing textiles to be sustainably discarded and reenter the biological cycle.

Project
Vank Cube

Designer
Anna Vonhausen

Material
Hemp and flax

Resource Type
Bio-based

Country of Origin
Poland

Manufacturer
Vank

Area of Use
Furniture design

Category
Reduced/cultivated

A modular framework of interlocking, bio-based components

Vank Cube is a modular furniture system made from a bio-composite material whose "building blocks" offer a future-proof solution for spaces with ever-evolving needs.

Developed using fast-growing, highly renewable flax and hemp fibers, Vank Cube embodies circular principles, reducing reliance on finite resources and promoting healthier ecosystems. Hemp can improve soil quality, protects against soil erosion, and absorbs 9–15 tons (8–13.5 tonnes) of carbon dioxide per hectare—comparable to that of a young forest—while maturing in just five months.

The cubes are designed for versatility, making the system well suited for temporary office spaces, exhibitions, trade shows, conferences, and workshops. They can be used in a variety of other settings as well, including co-working spaces or traditional workspace, offering adaptability to meet the changing needs of both the user and the organization.

The modular design allows multiple cubes to be assembled into desks, storage units, space dividers, or meeting tables, thanks to an interlocking system with reversible connectors and fasteners. Optional add-ons enable further customization, and include upholstered seat cushions, biomaterial side panels, and lightweight plywood tops, available in oak veneer or a range of color finishes.

Crafted from fast-growing, renewable flax and hemp fibers, Vank Cube's components can be assembled for use as a desk, storage unit, space divider, or meeting table thanks to its interlocking system.

Bioregional architecture shaped by material exploration

Lot 8, a 19th-century train-depot building, has been renovated to provide a new workspace for Atelier Luma. The space is a co-creation of the end user, Assemble, and BC architects & studies.

Project
Le Magasin Électrique

Designers
BC architects & studies, Assemble, and Atelier Luma

Materials
Rice straw, sunflower pith, timber, limestone dust, salt, and recovered roof tiles

Resource Type
Bio-based

Country of Origin
France

Area of Use
Architectural design

Category
Cultivated

Atelier Luma is a research design lab based in Arles, in southern France, where its primary workspace serves as a test bed for material research and architectural innovation. The architects worked with Atelier Luma to map the region's resources, industries, and waste streams. Together they identified nonextractive organically based and recycled materials to create a framework challenging the hegemony of industrial materials and move away from standardized modes of production. Dubbed "bioregional design," this approach focuses primarily on overlooked materials that are locally and readily available.

Nearly all the materials were sourced and processed within a 43-mile (70-km) radius of the site. The building's door handles are made from salt, harvested from nearby salt marshes. Furniture has been constructed from Japanese knotweed, and thermal insulation is composed of local rice straw from the Camargue region. Scaling up these material streams could have significant impacts, disrupting the traditional use of petrochemical-derived products. For example, according to researchers, just 5% of the rice straw produced in France could insulate every building in the country.

Many of the materials used in the renovation are made from the by-products and leftovers of other industries. The wall finishes and acoustic insulation

“Dubbed ‘bioregional design,’ the approach focuses primarily on overlooked materials that are locally and readily available.”

are made from agricultural waste, such as the fibers and mashed-up pith of sunflowers, and bathroom tiles were made using surplus clay from a local quarry. Internal walls are made using a rammed-earth construction that repurposes demolition debris and limestone residue from regional quarries.

Time was allowed for material choices to develop organically on-site, with no one material defining the architecture. This approach required the architects to engage in prototyping, to understand how novel materials would behave and react. This iterative process was crucial for developing their material knowledge and making informed decisions. The building itself is in a constant state of construction, envisioned as both everchanging and highly adaptable—effectively functioning as a prototype in its own right.

The finishes are not dictated by a predetermined aesthetic for the space but rather emerge organically, allowing the atelier's research streams to evolve and take shape.

Transforming used coffee grounds with parametric design tools

On one of Barcelona's tree-lined boulevards, the D·Origen Coffee Shop resides behind the Art Nouveau facade of Gaudí's Casa Calvet. Internally, the space is animated with bespoke 3D-printed furniture made from coffee grounds and PLA.

Project
D·Origen Coffee Shop

Designers
Lowpoly and Arturo Tedeschi

Material
Coffee grounds

Resource Type
Bio-based

Country of Origin
Spain

Manufacturer
Lowpoly

Area of Use
Furniture and fittings

Category
Recycled

Coffee is the one of the most popular drinks worldwide, with an estimated 400 billion cups consumed annually, according to Nestlé. Gianluca Pugliese from Lowpoly describes the hidden cost that is generated: approximately 60 million tons (54 million tonnes) of spent coffee grounds every year. Although much of these still end up in landfill, projects like the D·Origen Coffee Shop help demonstrate that used coffee grounds can be sustainably repurposed through forward-thinking initiatives.

The café's furniture is 3D-printed from a mix of coffee grounds and PLA, producing opulent forms inspired by the fluidity and richness of coffee. The shop features lamps, stools, and a back-lit counter, all made from Lowimpact, a biodegradable, petroleum-free material made with 98% organic content. Lowpoly partnered with Arturo Tedeschi to create the sculptural, expressive forms using artificial intelligence and parametric design software. Manufacturing relied on advanced large-format printers, industrial robotic arms, and specialized Rev3rd pellet extruders designed to handle materials with high organic content.

Reduced Materials

This category explores how thinking about material simplicity and resource efficiency from the inception of the design process helps facilitate ease of maintenance, future proofing, and end-of-life recovery.

Within the context of this category, *reducing* refers to minimizing excess through the streamlined use of materials and form, making products more resource-efficient. It explores projects that reduce complexity and complication by utilizing fewer constituent parts or materials, or through improved manufacturing processes—thereby reducing the overall carbon footprint and energy use involved in production. Designing for disassembly means reducing the need for dismantling or separating complex components, which can be time-consuming and costly. By minimizing the variety of materials used and simplifying connection points to nonpermanent fixings and fastenings, designers lower logistical and economic barriers to recycling and reuse; this includes phasing out chemical and irreversible fixings like glues and binders, helping ensure that materials can successfully reenter the manufacturing cycle.

Opposite and above: The interior of the Cork House by Matthew Barnett Howland, Dido Milne, and Oliver Wilton. The house's monolithic walls and corbelled roofs are made almost entirely from solid, load-bearing cork.

Mono-material thinking

Paradoxically, a well-executed, highly distilled, simple concept often involves innovation and technical complexity—particularly where one material is fulfilling the function of many. For example, the Mono [PA6] backpack by FREITAG (p. 50) is made entirely from nylon, despite each of the bag's 17 components having highly specific performance requirements and properties, from the buckles, carrying straps, and sewing thread to the water-repellent outer fabric.

This mono-material approach is also highlighted by projects like the Cork House by Matthew Barnett Howland, Dido Milne, and Oliver Wilton, which replaces what are typically multiple layers of a building's envelope with one material, cork (p. 98). To eliminate glue and mortar and achieve the high tolerances necessary for interlocking friction fits, every cork block has been precision-milled by a 5-axis CNC machine. While mono-material designs simplify end-of-life recovery, achieving this simplicity often requires highly technical procedures to ensure that the material meets diverse performance needs—in this case, sufficient airtightness and water resistance.

The projects, in some way, strive to eliminate an unnecessary complexity that can make a product difficult to use, look after, alter, repair, recycle, or reuse. Although the focus is generally on recycling and reuse, simplifying repair and maintenance is also an important strategy in extending a product's life. Box by Max Lamb (p. 24) is a collection of furniture pieces that can be easily repaired with readily available materials. Objects in the collection are made by combining cardboard with a paste made only of flour and water. Although the pieces are designed to be long-lasting, if signs of wear begin to show, the table and chairs can be patched up at home with some scraps of cardboard and homemade paste.

Vank Cube is a modular system that can be used to form desks, tables, sideboards, and storage units. Each module is made from flax and hemp.

The Mono [PA6] backpack by FREITAG. Made entirely from polyamide 6, or PA6 for short, commonly known as nylon.

Building in sustainability from the outset

These projects differ from the ones on reuse and recycling in that it focuses on the impact of decisions at the earliest stages of the design process, rather than addressing the management of materials following their initial life cycle. While reuse focuses on repurposing existing materials and recycling breaks them down to recover raw resources, this category emphasizes reducing material complexity at the design stage to limit waste from the outset. Recognizing that up to 80% of a product's environmental impact is determined during the design process, the projects highlight how intentional, mono-material strategies can drastically reduce waste and simplify future circular processes.

"The projects, in some way, strive to eliminate an unnecessary complexity that can make a product difficult to use, look after, alter, repair, recycle, or reuse."

Top: A chair from Max Lamb's cardboard series, Box. Above: A close-up of Tide by Stuart Haygarth, a chandelier made from objects found along the coastline of the U.K.

The Anthill Pavilion by Jeremy Waterfield, BC materials, and 45 students from RWTH Aachen University. The pavilion features compressed earth blocks and thatched reed cladding.

Project
The Tane Garden House

Designer
Atelier Tsuyoshi Tane Architects

Material
Thatch

Resource Type
Bio-based

Country of Origin
Germany

Area of Use
Architectural design

Category
Cultivated

Prioritizing nonextractive, aboveground materials

Located within the Vitra Campus, the Tane Garden House is built entirely from overground materials, as opposed to extractive, highly processed underground materials, which are accelerating climate change.

Japanese architect Tsuyoshi Tane is working to highlight the loss of collective knowledge in building with biomaterials. In many cultures, thatching skills have been passed down through generations, but the rise of cheaper synthetic substitutes now endangers the future of this craft. His design approach is rooted in researching traditional building methods, natural resources, and local contexts.

As traditional architecture contributes to the climate crisis by consuming underground resources, Tane opted for nonextractive, aboveground materials. Thatch cladding was selected for its insulative properties, as a low-impact alternative to conventional, petroleum-based insulation products. The materials are locally sourced, reducing transportation emissions and helping support local economies.

The building's base is constructed from granite that has travelled just 17 miles (28 km) from the quarry, and its structure is made of timber that has completed a journey of only 31 miles (50 km) from the Black Forest.

The building features a private meeting room, workshop, storage for the gardeners, and a publicly accessible rooftop—allowing guests to enjoy panoramic views of the garden and campus beyond.

Learning from vernacular building methods and local materials

Built by 45 architecture students of RWTH Aachen University in just two weeks, the HORST Anthill Pavilion in Asiat Park, Vilvoorde, Belgium, was created for the HORST Arts and Music Festival.

Project
HORST Anthill Pavilion

Designers
Jeremy Waterfield, BC materials, Democo, and students of Junior Professorship Act of Building RWTH Aachen University

Material
Compressed-earth blocks, reed, and timber

Resource Type
Bio-based

Country of Origin
Belgium

Manufacturers
Students of Junior Professorship Act of Building RWTH Aachen University, Raggers Rieten Daken (Dennis Raggers), Democo Mason team, and BC materials (Bregt Hoppenbrouwers)

Area of Use
Architectural design

Category
Cultivared/reduced

Designed to house artist Afrah Shafiq's video game installation *Where the Ants Go,* the pavilion uses compressed-earth blocks, which are protected from the weather by prefabricated reed-clad cassettes. As the pavilion was built by nonprofessional thatchers, the typically flush surface of traditional thatch was simplified into overlapping 10 × 20 ft (3 × 6 m) modules. This adaptation was made feasible by the steeper angle of the pavilion's facade compared to that of conventional thatched structures, reducing the risk of water infiltration. The outer facade's appearance is a pragmatic outcome, defined by the material used and the resources' natural length.

The substructure for the pavilion's outer facade is a combination of reclaimed and new, solid wood. This meant adapting the frame construction to what was salvaged from deconstruction sites in Brussels—primarily floor joists of up to 60 years old. Careful evaluation and adjustments were made to accommodate the natural deformations of the aged beams. Conceived as a process of design by making, this project demonstrates that embracing alternative materials and learning from vernacular building methods can inspire new, innovative architectural forms.

AS THE QUEEN LAYS DEEP IN HIBERNATION
0:47

The pavilion is the result of a playful, nonrestrictive design approach that provides space for decision-making while building and allows the teaching of material-specific craft, both in outcome and construction.

Balancing a low-carbon footprint with long-term durability

Through extensive research and investigation, Invisible Studio have designed a yoga studio utilizing rammed stone, developing a construction formula that is both high-performance and low-impact.

Project
Yoga Studio

Designer
Invisible Studio

Material
Rammed stone

Resource Type
Bio-based

Country of Origin
United Kingdom

Manufacturer
Ken Biggs Contractors

Area of Use
Architectural design

Category
Cultivated

Rammed-stone construction is often overlooked in the United Kingdom due to limited local expertise, but Invisible Studio has challenged this status quo with their innovative yoga studio at the Newt hotel in Somerset.

The building's outer walls are formed primarily of rammed stone, coming from the local Somerset quarry, giving the building a relationship to the landscape where it sits. The Hadspen limestone used in the mix gives the building a honey-colored hue characteristic of many older structures in Bath, while also helping reduce carbon emissions associated with the transportation and production of materials.

The stone was crushed into a fine aggregate and dry mixed on-site with water and a small amount of binder, composed mainly of lime and cement. Invisible Studio collaborated with the material science lab at the nearby University of Bath, testing different mixes to find a balance between low embodied carbon and long-lasting durability. This mix was poured bucket by bucket into a formwork, where it was compressed in sections, a process that is joyfully expressed by the walls' stratified, layered finish.

The building features a single large skylight, framing the sky and treetops, allowing light to cascade onto the walls and curved internal soffit, all lined with slatted beech.

A monolithic building system designed for disassembly

The Cork House presents a radically simple approach to building, exploring how expanded cork blocks can effectively perform all the functions of the building envelope.

Project
Cork House

Designers
Matthew Barnett Howland, Dido Milne, and Oliver Wilton

Material
Cork

Resource Type
Industrial waste/bio-based

Country of Origin
United Kingdom

Manufacturer
Matthew Barnett Howland

Area of Use
Architectural design

Category
Reduced

Modern buildings are often composed of various layers and components made from a myriad of materials and fixings, and are assembled by a team of skilled subcontractors in an elaborate process that produces an inherent complexity. This poses challenges at the end of the building's life cycle, with the difficulty of dissasembly often leading to destructive demolition.

By contrast, the Cork House presents a radically simple approach, inspired by low-tech dry-stone wall construction, a building method that uses no mortar. The monolithic system, intended for easy disassembly, explores the potential of expanded cork blocks to serve all functions of the building envelope—such as those traditionally managed by internal linings, membranes, insulation, structural elements, and external cladding.

The expanded cork used in the Cork House is a byproduct of cork forestry. Cork cultivation is low-impact; the outer bark of the tree is harvested every nine years or so without harming the tree itself. Low-grade cork, left over from this process, is granulated and heated in an autoclave to form blocks or billets known as expanded cork. During heating, the cork releases suberin, a natural binding material, meaning that the blocks are entirely plant-based.

“The tongue and groove system requires no glue or mortar, allowing for easy recovery of the cork at the end of the building’s life.”

All 1,268 cork blocks were precision-machined using a large-scale, 5-axis CNC milling machine to achieve the high tolerances necessary for interlocking friction fits. The tongue and groove system requires no glue or mortar, allowing for easy recovery of the cork at the end of the building’s life. While the concept is simple, the outcome resulted in complex block geometries that ensured airtightness and water resistance. The Lego-like blocks incorporate channels for water drainage and air-pressure neutralization slots to minimize the risk of wind-driven rain penetrating the dry joints in adverse weather conditions. In addition, numerous block types were developed to meet the challenges of corners, windows, eaves, and the many corbelled roof conditions.

Upon completion, the building was carbon-negative—meaning its cork and timber components had captured more atmospheric carbon than was emitted during its construction.

Prefabricated and locally sourced, natural building materials

According to Material Cultures, the Block House, a compact dwelling in Somerset, sequesters more carbon from the atmosphere than was produced during its construction.

Project
Block House

Designers
Material Cultures and Studio Abroad

Material
Softwood timber frame, larch cladding, hempcrete, and flint

Resource Type
Bio-based

Country of Origin
United Kingdom

Manufacturer
Hamptons

Area of Use
Architectural design

Category
Cultivated

The 970-sq-ft (90-m^2) Block House was built on the grounds of a private estate to replace the former Hermitage, lost in 2018 when a mature beech tree fell. The new timber-framed structure is designed to be low-impact and sensitive to its leafy surroundings, resting lightly on two flint trench foundations. Aboveground, a natural palette of building materials is used including locally sourced timber, hempcrete blocks, and wood-fiber insulation. The building is inherently low-carbon, however, the project budget did not allow for a whole life-cycle carbon analysis.

The lightweight hempcrete blocks enable ease of installation on-site, while also providing thermal mass and a natural rhythm and texture to the interior. Internally, the blocks are exposed, finished only with a clay-based paint that allows the wall assembly to breathe. Sourced from Buckinghamshire, the modules were prefabricated off-site to streamline construction and bypass the long curing times needed for the lime content of the blocks. Externally, the walls are clad in locally sourced larch and sit beneath a cedar-shingle roof. Oak colonnades support the roof's deep overhanging eaves on two sides, which shelter the building's dual verandas.

The sites' relationship to neighboring fields and mature trees shaped many aspects of Block House, including its structure and foundations.

A micro-scale circular system prompting local regeneration

Ishizaka Corp, a company specializing in recycling industrial waste, has added two small buildings to its headquarters that together demonstrate the feasibility of a self-contained circular system, recycling toilet wastewater.

Project
Toiletowa

Designers
Tono Mirai Architects

Material
Rammed earth

Resource Type
Bio-based and recycled materials

Country of Origin
Japan

Manufacturers
Sanwa, Velux, and Woodio

Area of Use
Architectural design

Category
Reduced/recycled

The buildings are located within a public park and constructed predominantly from biodegradable or recycled materials, avoiding the use of concrete. The first building, a public toilet, is enclosed by two semicircular, rammed-earth walls. The walls are built using NS-10, a recycled soil derived from salvaged gypsum boards and other construction materials, developed in collaboration with Ishizaka and IS Engineering.

A biotechnology called combined fermentation (EMBC), developed by Dr. Yasuhide Takashima, is used to treat the toilet wastewater. Tono Mirai says that "by allowing aerobic bacteria and anaerobic bacteria to co-propagate and coexist, changes in fermentation, decomposition, fermentation synthesis, and fusion occur, and the water is purified, and contains zero bacteria, malignant bacteria, [or] E. coli." Toilet wastewater is fully recycled and reused to water the land, promoting the growth of surrounding garden crops and regeneration of the soil. The second building, a nearby semicircular timber structure, contains wastewater tanks, emphasizing the recirculation process.

The interior walls and earthen floors of the toilet building are made of almost entirely recycled materials: soil, wood, and glass. The hand-wash and toilet bowls are made using recycled wood chips.

A hyper-local building arising directly from its context

Sited in Kabalagala, a central district of Kampala, the arts center will serve as a welcoming, inclusive hub for the artists' community in Uganda, providing three-month residencies alongside rentable studios and a gallery.

Project
32° East Arts Centre

Designers
New Makers Bureau and Localworks

Material
Rammed earth, sandstone, and wood

Resource Type
Bio-based

Country of Origin
Uganda

Manufacturer
Localworks

Area of Use
Architectural design

Category
Reused/cultivated

Designed in collaboration with local architect and contractor Localworks, the center is thoughtfully integrated into the cityscape, set atop a slope overlooking a patchwork of rooftops. The project exemplifies a low-carbon, "hyper-local" approach, utilizing materials that were sourced on the site: two redundant buildings and the clay-rich soil they stood on. The excavated soil, which gives the center its distinct pinkish hue, was used to form the rammed-earth walls and compressed-earth bricks. Materials salvaged from the demolished buildings were reused as aggregate, and wood initially used in the formwork for the rammed-earth walls was repurposed into roof shingles.

Locally sourced sandstone forms the base of the building, while cement lintels are cast against corrugated metal sheets, a ubiquitous local material, to introduce a horizontal rhythm to the facade. These elements, combined with textural brick patterns, create a dynamic and harmonious expression of local craft. Passive design strategies were prioritized over mechanized solutions to reduce solar gain, operating costs, and maintenance. A mono-pitched, overhanging roof provides shade and protects the thick earthen walls from driving rain; hit-and-miss brickwork at the clerestory level introduces dappled light and supports natural ventilation.

The material palette, sourced within a 25-miles (40-km) radius, is highlighted through distinct stratification, creating a horizontal rhythm on the exterior.

Each studio features a projecting steel-framed window seat with timber shutters, allowing occupants to effectively control thermal comfort.

Cultivated Materials

Reducing our dependence on fossil fuels is especially critical in Europe, where these resources are becoming increasingly scarce. By shifting to cultivated materials, we can foster more sustainable and resilient pathways.

Bio-based materials consist of a wide range of resources derived either fully or in part from renewable living systems, which include plants, animals, enzymes, microorganisms, and their by-products. They stand in opposition to extractive, non-renewable, fossil-based materials and include a range of compositions, from raw materials like timber and stone to composite materials made using cellulose, pectins, and other natural substances.

Above: CornWall by Circular Matters and Front Materials, a wall finishing material. Made using discarded corncobs.

Examining the construction industry, we have focused on viewing sustainability through the lens of passive house building, which prioritizes high energy efficiency to drastically reduce the need for heating and cooling. Principally this has required the use of high-performance petrochemical products like tapes, membranes, and expanded foams to greatly improve insulation, thereby reducing heat loss and improving airtightness. However, as design studio Material Cultures (pp. 104 and 148) points out, these fossil-based materials do not retain any trace of the social or ecological processes involved in their production, are incredibly hard to repair, and are susceptible to single points of failure. The designers also note that the materials often fail to achieve their intended lifespan—but endure in landfill.

The importance of bio-based materials

While environmentally conscious consumption often focuses on minimizing harm, bio-based materials present an opportunity to actively improve environmental outcomes.

By sequestering carbon, natural materials can positively impact the biosphere. Organic matter, such as soil, timber, and hemp, are all able to capture and store carbon from the atmosphere. Algae is one of the most efficient organisms for absorbing carbon through the process of photosynthesis, which it uses to create energy and grow.

While bio-based materials offer a promising solution, the growing demand for nonbiodegradable and petrochemical-based materials like plastic highlights the urgency of scaling these alternatives. Annual production of plastics is expected to double by 2050. By reusing and recycling our current stock, we can achieve greater efficiency and sustainability. These strategies alone, however, cannot address a demand for plastics that is expected to grow in line with a rising global population, projected to reach nearly 10 billion people by 2050. Charlotte McCurdy (p. 150) highlights that fossil-based products, even in circular systems, cannot overcome entropy—the natural breakdown and degradation of materials. This means that our goods and possessions tend inevitably toward disorganization, and as materials break, mix, and degrade they become less useful. Bio-based alternatives overcome this problem by harnessing free solar energy from outside their systems.

Monolithic mycelium benches designed by Carlo Ratti Associati in collaboration with Grown Bio.

newtab-22 collects discarded seashells from sources in the seafood industries. The shells are ground and processed into the material Sea Stone.

The challenges and opportunities

Recent research from Radboud University in the Netherlands found that biomaterials release 45% less carbon dioxide on average when compared with their fossil-based counterparts, although significant variations exist between materials.

To truly assess the environmental value of bio-based materials, we must consider how crops are managed. Key factors include the choice between monoculture practices and diversified farming, the use of pesticides and fertilizers, and how land use is impacted.

“If managed well, bio-based materials enable an ethical and cruelty-free method of production grounded in creating healthier environments.”

Top: Block House by Material Cultures uses hempcrete, larch, and flint. Above: Marine macro-algae have been developed by Charlotte McCurdy to produce bio-based sequins for a dress.

The Algae Sequin dress by Phillip Lim and Charlotte McCurdy. Algae bio-plastic sequins are sewn onto a SeaCell base mesh.

Projects like Totomoxle by Fernando Laposse (p. 122) address some of these questions. New economic policy established in the 1990s forced rural farmers in Mexico to standardize their crops and embrace the use of herbicides and fertilizers—a shift that quickly eroded and depleted the soil. Laposse worked directly with Mixtec farmers to again embrace ancient practices and reintroduce indigenous crops, creating a new, craft-based economy for farmers.

If managed well, bio-based materials enable an ethical and cruelty-free method of production grounded in creating healthier environments. This category looks at designers who are promoting the beauty and utility of bio-based materials, through technological innovation but also with age-old knowledge and indigenous practices, challenging our understanding of materials—their aesthetic potential, their life cycle, and the ways we value and care for them.

Empowering rural farmers through a craft-based economy

Totomoxtle is a veneer material developed in partnership with Mixtec farmers from southwest Mexico, who have seen their way of life threatened by globalization.

Project
Totomoxtle

Designer
Fernando Laposse

Material
Corn husks

Resource Type
Agricultural waste/bio-based

Country of Origin
Mexico

Manufacturer
Local Mixtec production

Area of Use
Product and furniture design

Category
Cultivated

Totomoxtle, which means "corn husk" in the Nahuatl language, uses what would otherwise be considered a waste product to revive employment in rural Mexico.

When the 1994 North American Free Trade Agreement put small Mexican farmers into direct competition with US industrial corn producers, the only way to compete was by embracing the use of hybrid seeds, herbicides, and fertilizers—abandoning ancient Mesoamerican methods that kept the land fertile by planting the corn with beans and squash. This shift led to land degradation, loss of native seeds, fewer jobs, and a surge in migration from around the country—migration to the United States rose by 79% between 1994 and 2000.

Totomoxtle has created a craft-based economy for local farmers, helping them keep planting indigenous heirloom corn to provide material; its rich color palette is a visual reminder of the biodiversity we are losing. To make the material, husks are ironed flat, glued onto a textile backing, and laser-cut into puzzle-like pieces that can be assembled as veneer for decorative wall coverings and furniture pieces.

Where industrial agriculture rewards standardization, Totomoxtle favors the vivid and diverse color spectrum of heirloom corn—from soft pinks and warm ambers to deep purples.

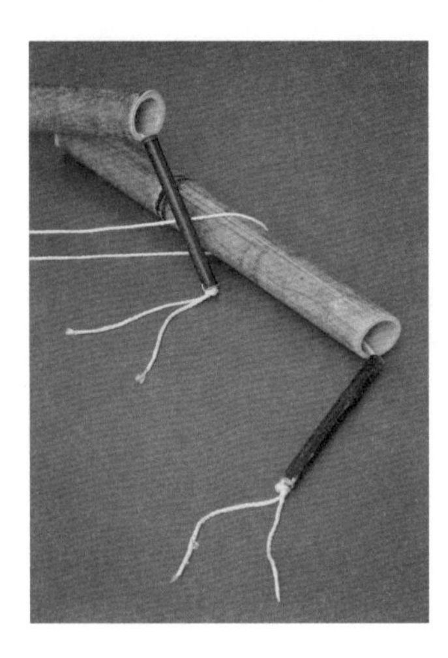

A modern revival of traditional Japanese craft

Bamboo has been used as a building material for centuries. In Kochi, Japan, however, where forests cover 84% of the land, bamboo is valued both as a traditional craft material and a crucial resource for the future.

Project
Soba Collection

Designer
Diez Office (Stefan Diez and Arthur Desmet)

Material
Bamboo

Resource Type
Bio-based

Country of Origin
Japan

Manufacturer
Taketora

Area of Use
Furniture design

Category
Cultivated

In the West, Bamboo has long been called "the poor man's timber," perhaps because of its abundance in developing tropical countries. This characterization reflects a perspective dismissive of these regions, one extended even to building materials—and overlooks the material's true value. Bamboo sequesters carbon, requires no pesticides or fertilizers, can regenerate degraded land, and help with soil erosion. It is also strong, lightweight, and highly renewable.

Stefan Diez worked directly with bamboo master Yoshihiro Yamagishi to develop the Soba Collection, exploring bamboo in its natural state as part of Japan Creative's initiative to revive time-honored Japanese crafts in a contemporary context. The minimalist collection features a flat-pack bamboo bench and a pair of trestle table legs, assembled without any tools or Western fittings. Bamboo shears easily due to the direction of its fibers, creating a risk of splitting the material when it is bolted through. Instead, Kevlar rope is threaded through eyelets and tied tight to hold the canes together.

Incisions made in the trestles keep the legs connected by a strip of bamboo, allowing them to flex and enabling them to be flat-packed, with easy assembly.

A nonwoven textile made from pineapple leaf fibers

Once pineapples have been harvested for their fruit, the plant's leaves are often burned. Globally, it is estimated that 44,000 tons (40,000 tonnes) of pine-apple "waste" are generated each year.

Project
Piñatex × Cat Footwear

Designers
Cat Footwear and Ananas Anam

Material
Pineapple leaves

Resource Type
Agricultural waste/bio-based

Country of Origin
Philippines

Manufacturer
Cat Footwear and Ananas Anam

Area of Use
Fashion

Category
Recycled/cultivated

Cat Footwear has partnered with Ananas Anam to reinvent the Intruder and Trespass boots using a leather alternative made from pineapple leaves.

The material repurposes agricultural waste from pineapple plantations in the Philippines by extracting fine cellulose fibers from the leaves and felting them into a nonwoven fabric that can be used to make clothing, shoes, car seats, and more. When this resource is used instead of burned, each linear meter of material prevents the equivalent of 26 lb (12 kg) of carbon dioxide from being released. In 2024, they saved 3,320 tons (3,013 tonnes) of pineapple leaves from becoming waste.

The product is 100% vegan and can be cut like leather without the wastage caused by irregularly shaped cowhides. While it won't completely replace leather, Piñatex offers a sustainable alternative that reduces the leather industry's negative impact on the environment, ecosystems, and human health. Traditional leather tanning uses harmful chemicals that can pollute soil and water sources. According to animal rights organization PETA, a study found that leather-tannery workers in Sweden and Italy face cancer risks 20% to 50% higher than expected.

CAT

Repositioning thousands of tons of European flax fibers

After becoming interested in flax, Christien Meindertsma purchased an entire harvest of the crop, equaling 11 tons (10 tonnes) of the raw material, from a farmer in the Netherlands.

Project
Flax Chair

Designer
Christien Meindertsma

Material
Flax

Resource Type
Bio-based

Country of Origin
Netherlands

Manufacturer
Enkev

Area of Use
Furniture design

Category
Cultivated

The initial appeal of flax stemmed from its low-impact cultivation, as it requires very little water and thrives in Northern Europe. When Meindertsma learned that much of the crop is now exported to China, she became interested in bringing linen production back to Europe.

Traditionally, flax's long fibers are used for linen; the short fibers are discarded or used for insulation. After weaving 2,600 ft (800 m) of linen, Meindertsma faced a challenge—what to do with 4.4 tons (4 tonnes) of the inferior short fibers. She partnered with specialist manufacturer Enkev to combine the short fibers with polylactic acid, a resin-like, biomaterial glue made from sugarcane or cornstarch, creating a new, felt-like material.

The flax chair is 100% biodegradable and designed to be made from a single sheet of the composite material, which combines four layers of woven flax fabric and five layers of the newly developed flax felt. The seat and back are cut from the center of the sheet, with the leftover shape forming the legs—leaving zero waste. The pieces are heat-molded, and incredibly, need no additional strengthening.

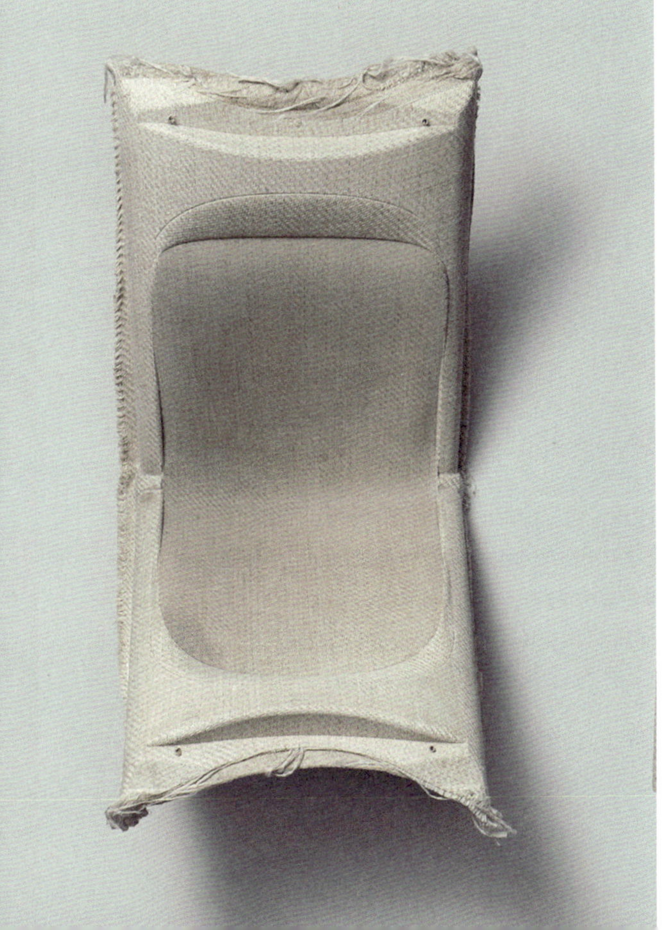

Harnessing Sicily's orange-industry waste

Orange Fibre is an Italian clothing manufacturer creating circular, silk-like fabrics from the by-product of citrus juice production.

Project
The Colours of Shadow at Midday

Designer
Mario Trimarchi

Material
Orange peels

Resource Type
Industrial waste/bio-based

Country of Origin
Italy

Manufacturers
Orange Fibre and Salvatore Ferragamo

Area of Use
Fashion

Category
Recycled/cultivated

Italy is one of the European Union's main orange producers, handling around 770,000 tons (700,000 tonnes) per year. The resulting orange-peel waste, accumulating mainly through orange juice production, presents significant environmental and economic challenges, with approximately 40%–60% of the mass of a typical orange being discarded as waste. Traditional disposal methods such as composting and animal feed lack financial appeal, while landfilling and burning lead to the creation of greenhouse gases. Globally, around 120 million tons (108 million tonnes) of citrus waste are produced annually.

Orange Fibre is a Sicilian company that has created a sustainable, silk-like fabric from the by-product of orange processing. The leftovers of the orange are collected, and the cellulose is extracted, turned into fibers, spun into yarn, and finally weaved into fabric.

Salvatore Ferragamo launched a daily-wear capsule with the fabric, featuring a print design from Mario Trimarchi. Originally drawing the design by hand, Trimarchi took inspiration from shadows cast on Mediterranean fruit and vegetation.

Salvatore Ferragamo

A renewable material made from discarded seashells

Sea Stone is a low-impact, terrazzo-like material made from the discarded oyster, mussel, and abalone shells that accumulate during seafood processing.

Project
Sea Stone

Designer
newtab-22

Material
Seashells

Resource Type
Industrial waste/bio-based

Country of Origin
South Korea/United Kingdom

Manufacturer
newtab-22

Area of Use
Product design

Category
Cultivated

According to newtab-22, trades such as commercial fishing and aquaculture are responsible for discarding around 7 million tons (6 million tonnes) of seashells every year. Some of the waste can be used to produce fertilizer; however, the greatest part is not managed effectively but is left to pile up along beaches or in landfills.

The designers collect the shells directly from these industries, cleaning and sorting them before grinding them down into fragments and mixing with natural, nontoxic binders. This amalgam is left to set and harden in a mold, creating a stone-like material that has been used to create tiles, vases, and tabletops. This entire process is carried out by hand to minimize the use of heat and electricity, contributing to a low-impact output.

Like limestone, a key ingredient in cement production, the shells are composed primarily of calcium carbonate. Given the shared properties of the materials, newtab-22 aims to position Sea Stone as an alternative to concrete for small-scale crafts. For larger works, the necessary heating process, which is similar to that of cement, is too carbon-intensive to be viable.

Natural variations in the texture and color of shell fragments make each piece of Sea Stone unique. The material has been crafted into decorative tiles, tabletops, plinths, and vases.

Project
The Modern Seaweed House

Designer
Vandkunsten

Material
Eelgrass

Resource Type
Bio-based

Country of Origin
Denmark

Manufacturer
Vandkunsten

Area of Use
Architectural design

Category
Cultivated

A contemporary approach to a traditional building craft

Seaweed-clad houses have been part of Læsø's rich history for centuries, with hundreds of them once lining the Danish island, but following a fungal disease outbreak in the 1920s, only around 20 now remain.

Læsø's seaweed houses can be traced back to the Middle Ages and the then-booming salt industry; as trees were burned to feed the salt kilns, residents were forced to look for alternative building materials. The Modern Seaweed House is a contemporary take on the island's vernacular, and an attempt to preserve its traditional building craft.

Historically, seaweed was stacked high on the roof, but the Modern Seaweed House is more contemporary in its expression. The "seaweed"—specifically, eelgrass—clads the exterior in tightly stuffed woollen nets, and is also packed into cassettes within the wall, floor, and roof construction, providing insulating properties comparable to those of mineral wool.

Once the eelgrass washes ashore, it is harvested and left to sun-dry before being pressed into bales. According to the architects, the resulting material is nontoxic, rot- and pest-resistant, and has an expected life of 150 years. The eelgrass also captures carbon dioxide—with lifecycle analysis calculating that the building has sequestered around 10 tons (8,500 kg) more carbon dioxide than was emitted during its construction.

Project
Savian by BioFluff

Designer
Ganni

Materials
Nettle, hemp, and flax

Resource Type
Agricultural waste/bio-based

Country of Origin
Denmark

Manufacturers
BioFluff and Ganni

Area of Use
Fashion

Category
Cultivated

An ethical alternative to animal and synthetic fur

According to the manufacturer, Savian is the first plant-based alternative to fur, fleece, and shearling, representing a radical departure from conventional and faux fur, which, respectively, rely on animal suffering and on plastic and petrochemicals.

The material is 100% plant-based, plastic-free, and GMO-free. It is made from a mixture of nettle, hemp, and flax harvested from renewable sources and agricultural waste streams in Europe. Manufacturing takes place in Italy, with a patented process using plant-derived enzymes.

The "hairs" of the "fur" are plant fibers, which do not need to be processed into yarn, making the energy-intensive yarn-spinning process unnecessary. Although a full life-cycle analysis is yet to come, BioFluff estimates a 50% reduction in greenhouse gas emissions during the manufacturing process as compared to conventional, plastic faux fur. In comparison with the production of real fur, the manufacturer estimates a 90% reduction in emissions, given the immense impact of fur farming, as well as the use of toxic chemicals in fur processing.

Ganni has collaborated with BioFluff, using the material to produce a version of the company's classic Bou bag, which was unveiled at Copenhagen Fashion Week 2024. The bag's Savian body is dyed either pink or black with mineral pigments and finished with a woven handle made of recycled leather.

Farming regenerative material from agro-industrial fruit waste

To support its goal of cutting carbon emissions in half by 2027 (from a 2021 baseline), fashion brand Ganni has phased out leather, instead using Celium, a vegan, plastic-free alternative, to create a crop top, miniskirt, and their signature Bou bag.

Project
Celium by Polybion

Designer
Ganni

Material
Cellulose

Resource Type
Agricultural waste/bio-based

Country of Origin
Mexico

Manufacturers
Polybion and Ganni

Area of Use
Fashion

Category
Cultivated

Celium is grown by feeding bacteria agro-industrial fruit waste through a fermentation process, during which the cells self-organize and create a cellulose structure as a metabolic by-product. This structure is then stabilized to achieve Celium's high-performance characteristics, resulting in a material that can be dyed, embossed, and tanned using chromium-free formulations.

The journey from raw material to a refined, leather-like finish is completed within a 30-mile (48-km) radius of Ganni's production plant in Guanajuato, Mexico. Utilizing existing infrastructure, such as large-scale fermentation equipment already present in tanneries, further increases the feasibility of scaling production.

Celium boasts a much smaller carbon footprint than that of traditional and synthetic leathers alike, benefiting from a fully solar-powered facility and upcycling waste that would otherwise go to landfill and produce methane. According to Polybion, third-party analysis indicates that the production process emits 99% less carbon dioxide than most animal leather production, and 61% less than production of most plastic-based leather alternatives.

GANNI

Project
Woodland Goods

Designer
Material Cultures

Material
Bark, natural lignin glue, and pine needles

Resource Type
Bio-based

Country of Origin
United Kingdom

Manufacturer
Erthly

Area of Use
Sheet material for design and construction

Category
Cultivated

Utilizing woodland resources for a postcarbon future

Material Cultures explores overlooked building materials from Britain's woodlands, demonstrating how sustainable alternatives beyond conventional processed wood can support the construction and furniture industry.

Britain's forests have suffered dramatic losses, with more ancient woodland destroyed in the 40 years after the Second World War than in the previous four centuries, as noted by conservationist Isabella Tree. Through their research, Material Cultures identified a heavy reliance on coniferous trees for the country's construction industry, a practice which can lead to a monoculture forestry that threatens biodiversity and native species.

Drawing from Indigenous practices and reexamining underused woodland resources—bark, natural lignin glue, and pine needles—Material Cultures aims to find alternatives to commonly used sheet materials and cladding composites that are free from plastics and harmful glues. The solid, naturally waterproof sheets they have developed are made by layering strips of sequoia, pine, and birch bark in alternating directions, followed by compression and heating. This process activates the lignin in the bark, which acts like a natural glue and binds the sheet material components together.

The studio also tested spruce-bark chips and Scots pine needles from the forest floor—resources that are often underused or considered waste in the lumber industry. Although these materials did not respond to heat and pressure alone, the studio has explored the potential of bio-resins for developing composite sheets.

Offering alternatives to conventional polyester sequins

Every year, millions of sequined garments are purchased for festive occasions alone, many of which are quickly discarded, releasing microplastics into the world's oceans.

Project
Algae Sequin Dress

Designers
Phillip Lim and Charlotte McCurdy

Material
Algae

Resource Type
Bio-based

Country of Origin
United States

Manufacturer
Pyratex

Area of Use
Fashion

Category
Cultivated

Phillip Lim and Charlotte McCurdy teamed up as part of an incubator project seeking to pair high-profile fashion designers with experts in sustainability. McCurdy had previously developed a thin bioplastic from macro-algae, which she used to produce a raincoat. Together, the duo recognized the potential in a new application for the material: replacing conventional polyester or vinyl sequins, which cause significant environmental harm.

The bio-based sequins are produced through a process of casting heated algae into glass molds, which transfer a reflective finish onto the resulting curve-shaped pieces. Mineral pigments were added in the casting process to achieve a luminous shade of green that shines with a soft, wavering light. The sequins are sewn onto a base mesh of cellulose fibers called SeaCell, a biodegradable textile derived from seaweed and bamboo.

Although the dress is not commercially available beyond custom orders, McCurdy envisions these innovative materials driving an impact at scale through the research and development they encourage simply by their presence in the market.

A carbon-positive wall finish made from agricultural waste

CornWall is a fully biodegradable interior wall-cladding product derived predominantly from corncobs—one of the world's most abundant agricultural by-products.

Project
CornWall

Designer
Circular Matters and Front Materials

Material
Corncobs

Resource Type
Agricultural waste/bio-based

Country of Origin
Netherlands

Manufacturer
Circular Matters and Front Materials

Area of Use
Interior wall cladding

Category
Cultivated

CornWall offers a sustainable alternative to traditional ceramic wall tiles and plastic laminates, available in both tile and sheet formats. Composed of 99.5% plant-based materials, it contains corncobs from Western Europe and other agricultural by-products. The remaining 0.5% consists of non-organically based color pigments that are biodegradable.

The material is formed into sheets in a low-energy pressing process, at a modest heat of 250–300 °F (120–150 °C). This process is powered entirely by renewable energy, largely from solar panels installed on the production facility's roof. Impressively, the production of CornWall is carbon-positive, with the material sequestering more carbon dioxide than is emitted during manufacturing.

The tiles are designed with a mechanical fixing system, enabling easy demounting for reuse or return to the manufacturer for cleaning and recycling. This eliminates the need for glues or adhesives, ensuring that all raw materials can be fully recovered and reused or recycled, aligning with circular design principles.

Sourced primarily in Western Europe from the cores of corncobs, this organic waste is widely available and otherwise destined for burning as biomass, fermentation, or left to rot.

Project
nat-2 Sneaker Lines

Designer
Sebastian Thies

Materials
Orange scraps, cacti, Bananatex, rose petals, moss, leaves, cork, and rubber

Resource Type
Industrial waste/bio-based

Country of Origin
Italy/Germany

Manufacturers
K&T and Ohoskin

Area of Use
Fashion

Category
Cultivated

A sneaker range that that creates its own virtuous circular economy

Throughout its collections, nat-2 has developed innovative materials, producing shoes from vegan products and industrial waste streams, such as recycled Bubble Wrap, recovered aluminium foil, and leftover milk from the dairy industry.

Sleek Low Fruit Orange Sun, part of nat-2's fruit collection, is a 100% vegan sneaker produced in cooperation with Italian textile maker Ohoskin from recycled orange scraps and cactus pear—otherwise referred to as the prickly pear, the Indian fig, or in Sicily, *bastardoni*—"big bastards."

Prickly pear cacti are largely grown for their milk for use in the beauty industry, but their skin is often perceived as waste, which can bring high environmental and economic costs. This is a particular issue in Italy as the world's second-largest producer of cacti, with 90% of production taking place in Sicily.

The sneaker is produced under ethical conditions by a small family-run manufacturing firm in Italy. The shoe features a removable cork insole, a bioceramic lining, and a durable rubber outsole. Another sneaker in the fruit collection, the Sleek Banana, is produced in partnership with Green Product Award and Bananatex and is handmade in Italy using real honey-impregnated Bananatex. The collection offers a cruelty-free and sustainable alternative to traditional leather sneakers.

The sneaker features a removable cork insole, a bioceramic lining, and a durable rubber outsole. The nat-2 shoes are produced under ethical conditions by a small family-run manufacturing firm in Italy.

MADE IN

Degrowth and the slowing down of furniture production

Made from fallen ash wood, naturally dyed with blue-flower tea waste and inspired by the gentle, fluid motion of water ripples, the Blueprint collection challenges the notion of trend-driven, fast furniture.

Project
The Blueprint Collection

Designer
Studio Kloak

Material
Fallen ash wood and butterfly pea flower (blue) tea

Resource Type
Bio-based

Country of Origin
United States

Manufacturer
Studio Kloak

Area of Use
Furniture design

Category
Reduced/cultivated

In the United States, shifts in consumer behavior have been shaped by the declining quality of mass-market furniture. During the mid-20th century, much of the furniture produced in the United States was crafted from plywood or solid, high-quality hardwood. By the 1970s and 1980s, however, the market saw an influx of engineered wood products, such as oriented strand board (OSB) and medium-density fiberboard (MDF).

Over time, furniture became cheaper and less durable, and repairs were disincentivized. Today, new sofas can be purchased from major online marketplaces for as little as $25. Trend-driven consumption has fostered the perception that furniture is disposable, designed to last only a few years—overlooking the environmental cost and the toll on workers across the supply chain.

Many furniture producers prize manufacturability above all, churning out large volumes of homogenous pieces, but Studio Kloak takes a different route. Its Blueprint Collection is produced through labor-intensive hand carving—a process difficult to replicate at scale—combined with a faded dye rather than block color, which favors standardization. Studio Kloak also partners with the city of Evanston, Illinois, to source local wood from trees that would otherwise become mulch, and repurposes waste tea leaves from a local Thai restaurant for the dye.

Optimizing carbon capture in dense urban environments

Photo.Synth.Etica provides an incubator for micro-algae, streamlining the process of carbon sequestering. These "urban curtains" are designed to remove pollution and purify air in dense urban environments.

Project
Photo.Synth.Etica

Designer
ecoLogicStudio

Material
Algae

Resource Type
Bio-based

Country of Origin
United Kingdom

Manufacturer
ecoLogicStudio

Area of Use
Architectural design

Category
Cultivated

Modernity advocates for the removal of bacteria and microorganisms from our urban realms, with architecture acting as an instrument to control and manage the natural world. Now ecoLogicStudio is re-conceptualizing our cities in favor of a bio-digital paradigm, in which architecture and nature work in symbiosis.

The London-based studio has designed "urban curtains"—optimized habitats for micro-algae that can be applied to building facades. As unfiltered air enters through the bottom of the curtain-like vessels, micro-algae integrated within capture and trap carbon dioxide molecules and air pollutants, which they in turn used to grow into biomass through the process of photosynthesis. The biomass can be harvested and converted to energy or used to make a bioplastic (the same material from which the facade units are made). To conclude the process, clean oxygen is released from the top of each module.

Photo.Synth.Etica modules in Dublin, Ireland capture approximately 2 lb (1 kg) of carbon dioxide per day, equivalent to the capture of 20 large trees. The bioreactors' footprint is much more efficient, and allows greater flexibility in crowded urban contexts.

AN CHLÓLANN
Photo.Synth.Etica
a bio-digital solution to global carbon sequestration
by ecoLogicStudio

Recycling fruit peels and coffee grounds into household goods

Krill Design has developed a fully biodegradable, natural material from waste orange peels, lemon peels, and coffee grounds—helping to address waste management challenges.

Project
Rekrill

Designer
Krill Design

Material
Orange peels, lemon peels, and coffee grounds

Resource Type
Food waste

Country of Origin
Italy

Manufacturer
Krill Design

Area of Use
Product design

Category
Recycled/cultivated

The designers have developed a patented process to recycle industrial organic waste such as peels, shells, seeds, and coffee grounds. The peels and other organic scraps are dried, ground to a fine powder, and mixed with a PHA biopolymer derived from bacteria. This mixture is used to produce pellets, which can be extruded as filament for the manufacture of various 3D-printed products. Rekrill is 100% vegan and estimated to save 95% carbon dioxide compared with conventional plastic filament.

Krill's Ohmie lamp is 3D-printed using orange peels from southern Italy—supporting a local supply and production chain. It takes only several hours to 3D-print and can be broken down and composted at the end of its life, for disposal with other household organic waste. While bioplastics are a rapidly growing sector offering hopeful solutions, they currently represent less than 1% of the about 400 million tons (362 million tonnes) of plastic produced annually. The company is now examining how they can scale Rekrill with injection molding techniques and have begun selling the filament directly to consumers, enabling them to locally 3D-print custom designs with the material.

I AM AN
ORANGE

I AM A
LEMON

ALEA
ISSUE
VISCOSE

Krill Design repurpose by-products from food and beverage companies operating in Italy. Because waste from these businesses is highly regulated, it arrives exceptionally clean and standardized.

An acoustic material that reflects Taiwan's rich indigenous practices

By incorporating local artisanal modes of production, Studio Flaer has developed an acoustic textile that utilizes circular materials and preserves cultural heritage.

Project
Indigo

Designers
Studio Flaer and the National Taiwan Craft Research and Development Institute

Materials
Paper mulberry and banana plant fibers, bamboo, and indigo

Resource Type
Bio-based

Country of Origin
Taiwan

Manufacturer
Local artisans

Area of Use
Acoustic panels

Category
Cultivated

Indigo is an acoustic in-room structure that represents elements of Taiwan's cultural heritage. The panels' surfaces embed the language of local patterns, customs, and traditions—from basket weaving and tile making, to the holistic practice of walking barefoot on cobbled reflexology paths.

Through collaboration with local artisans and material researchers at the National Taiwan Craft Research and Development Institute, heritage crafts are translated into sound-absorbing panels made entirely from organic banana and paper mulberry fibers. Bamboo is used to frame the lightweight panels, while the indigo plant gives them their blue color and name.

Since the banana is a flowering plant, once the fruits have been harvested the plant dies and becomes a resource with various potential uses. Paper mulberry, a native but invasive plant that spreads quickly and supersedes other local flora, is another underused resource. Although its fibers can be used in paper making, the plant is often cut up and discarded.

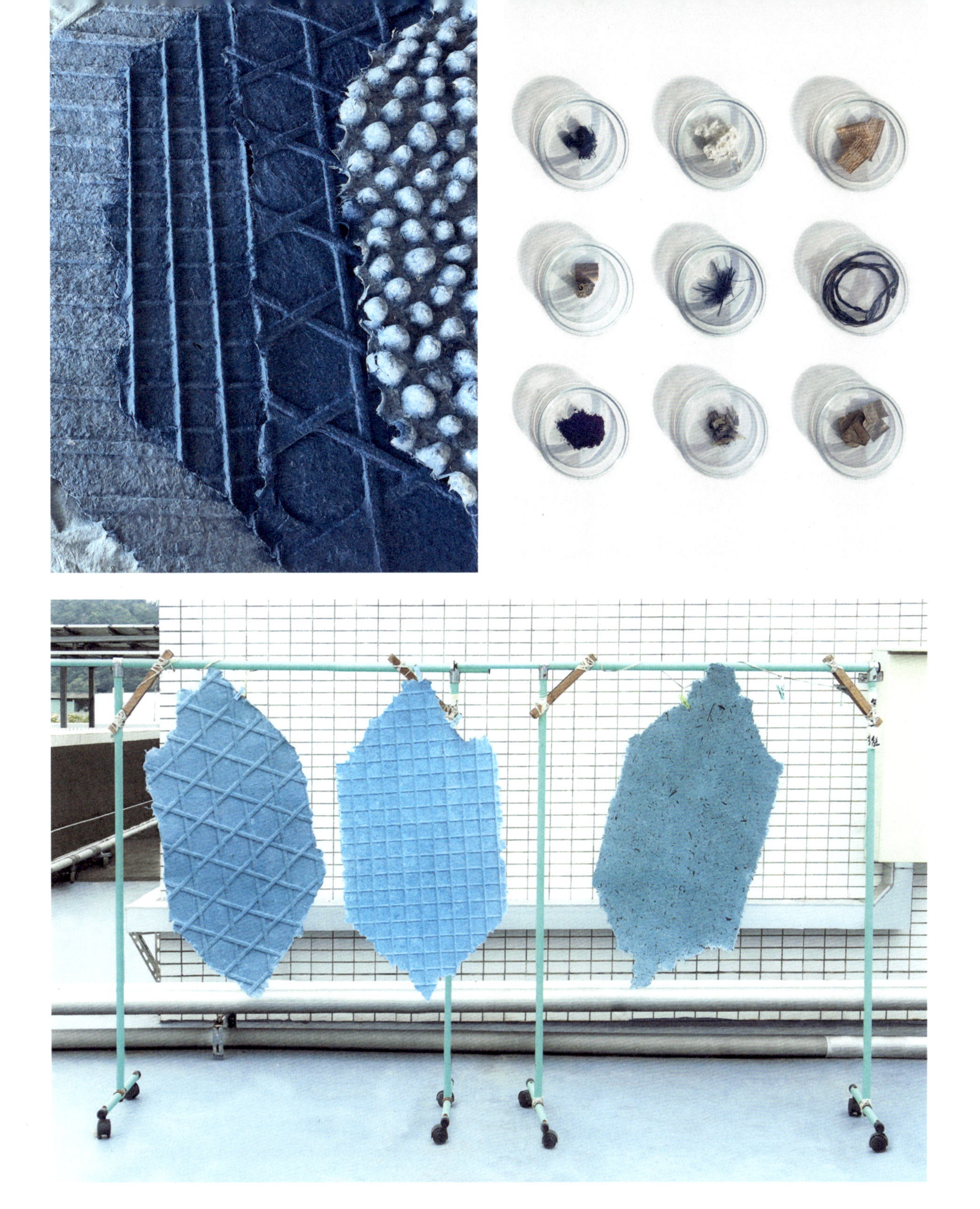

Working in co-creation with Taiwan's indigenous tribes and the National Taiwan Craft and Research Institute, Studio Flaer developed plant-based, circular materials.

Growing furniture from living matter and organic waste streams

By cultivating mycelium to produce flowerpots, side tables, and bench seating, companies like Grown Bio are challenging conventional notions of beauty—and asking how biomaterials might engender new aesthetic sensibilities.

Project
Mycelium Furniture

Designers
Grown Bio, Danielle Trofe, and Carlo Ratti Associati

Material
Mycelium and hemp

Resource Type
Agricultural waste/bio-based

Country of Origin
Netherlands

Manufacturer
Grown Bio

Area of Use
Furniture and product design

Category
Cultivated

Grown Bio is a Dutch biotechnology company that grows furniture and packaging materials. The company initially set out to develop fully biodegradable mycelium packaging to replace polystyrene, which is difficult to break down and contains diminishing resources. Currently, the company sells a range of products, including building materials and interior furnishings.

The products are made of only two ingredients: local agricultural waste and mycelium—the underground root network of fungi. First, these ingredients are poured into a recycled plastic mold, which can be reused many times. In a controlled, humid environment, the mycelium then feeds on the waste substrate, which is often hemp. As the mycelium grows, it binds the biomass together, taking the shape of the mold. To stop the material from continuing to grow, the product is then placed in a drying chamber for two days.

Mycelium products are lightweight, compostable, free of volatile organic compounds (VOCs), and said to have fire-retardant properties. Grown Bio has collaborated with designer Danielle Trofe on their MushLume lighting pendants and Carlo Ratti Associati to design a collection of monolithic benches.

THE
CANALS

OCEANS
GROWN.bio

Grown rather than manufactured, Grown Bio embraces a natural, organic aesthetic and offers an earthly alternative that can be composted at home once it reaches the end of its life.

A fully biodegradable fruit-leather bag made of plant-based fibers

Two different industrial by-products are used to manufacture Sonnet155—short cellulose fibers rejected by the textile industry and a plant-based fiber left over from fruit juice production, called pectin.

Project
Sonnet155

Designers
Johanna Hehemeyer-Cürten and Lobke Beckfeld

Material
Cellulose and pectin

Resource Type
Industrial waste/bio-based

Country of Origin
Germany

Manufacturers
Johanna Hehemeyer-Cürten and Lobke Beckfeld

Area of Use
Fashion

Category
Cultivated

Cellulose fibers discarded by textile factories because of their short length are used to make the bag's sheet material. These fibers, which are under ¼ inch (5 mm) long, are combined with pectin, which is extracted from waste fruit peels and used as a binding agent. The mixture is combined with warm water, left in a mold to cure for up to five days, and then the bag is sewn into shape. Natural pigments are used to add color; the materiality of the mold itself determines if the finish is matte or glossy. Every bag has a unique patina due to variations in the multiple characteristics of the fibers.

Designed to be temporary, the product has a lifespan comparable to that of a disposable paper bag, and can be composted or recycled after signs of wear. The whole life cycle of the materials has been considered, allowing for the two main ingredients to be easily separated and recovered. The material can be dissolved in warm water and recast to create another bag. Alternatively, the cellulose can be filtered out with a sieve and reused, and the pectin repurposed as plant food.

The lengths and density of cellulose fibers, as well as the overall cellulose content, inform the texture, opacity, and durability of the material.

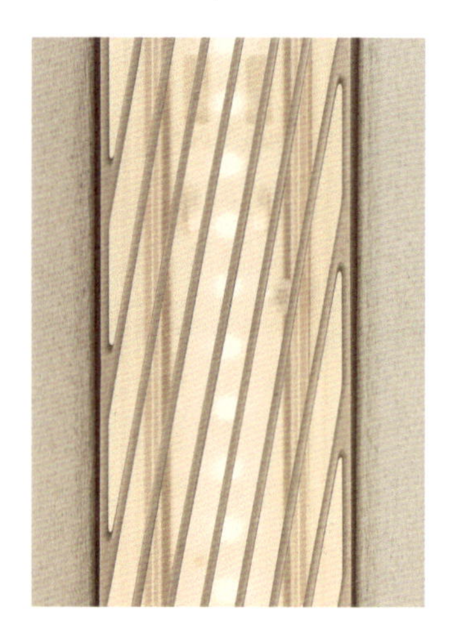

Project
Superdupertube

Designer
Snøhetta

Material
Hemp, sugarcane, and wood cellulose

Resource Type
Bio-based

Country of Origin
Sweden

Manufacturer
ateljé Lyktan

Area of Use
Lighting design

Category
Reduced/cultivated

Updating a 1970s design classic through extensive material research

Snøhetta and lighting brand ateljé Lyktan have reimagined the 1970s Supertube lamp—an office light originally made from extruded aluminum—with an updated design that cuts the lamp's carbon footprint by using a base of extruded hemp and sugarcane bioplastic.

Initially, the design team experimented with pine cones and coffee grounds, but these proved too fragile due to their short fibers. The team subsequently developed a robust, hemp-based composite mixed with wood cellulose, minerals, and a bioplastic derived from sugarcane. This innovative material reduces carbon dioxide emissions by 73% and weight by 53% compared to the original aluminum design.

The lamp's body is extruded from this material, honoring the manufacturing process of the 1970s classic. Additional components, such as injection-molded side covers and glare-reducing louvres, are assembled without glue and with a minimum of screws—enabling easy disassembly and repair. At the end of the lamp's life cycle, the parts can be returned to the manufacturer for industrial composting or recycled into pellets to make more lamps.

Because 70–80% of a luminaire's climate impact occurs during the use phase, the design incorporates smart lighting systems, including motion sensors to detect presence. These systems reduce energy consumption by up to 80% by adjusting the lighting based on room occupancy.

SHOW ME THE
HARMONY

Reused Materials

By reusing materials that have lost their perceived value, we slow resource loops and extend their useful lives.

The long-term aim is to shift from a linear, "take-make-waste" model to circular systems that feed materials back into the production cycle—keeping materials and products in the system at their highest-value state for as long as possible. After repairing and maintaining, reusing is seen as the next-best way to keep nonbiodegradable materials in the system.

Above: A table setting found in the Circus Canteen by Multitude of Sins, a restaurant in Bangalore producing zero food waste.

Reuse in construction presents challenges, particularly during demolition, when significant time and care are needed to dismantle and catalog materials. Logistical hurdles—such as storage, transportation, and obtaining warranties or certifications—add further complexity. The assumption that a material is not worth the time and effort to salvage, however, often ignores the object's embodied carbon. "Embodied carbon" refers to the emissions generated throughout a material's life cycle—from extraction and production to transport, construction, and disposal. It can account for 35–50% of a building's total emissions, depending on its typology. While efforts have focused on reducing operational carbon through renewable energy, progress on lowering embodied emissions has moved much more slowly.

Going against industry norms

Reuse requires going against long-established norms and competing against industries that have enjoyed the benefits of time and substantial investment. Industries, including fast fashion and tech, drive massive waste generation, and turn consumption into a political and cultural act. The Ellen MacArthur Foundation, an environmental nonprofit organization, reports that the equivalent of one garbage truck full of clothing is burned or goes to landfill every second. By contrast, slow fashion and secondhand clothing limit waste and reduce the need for resource-heavy materials like virgin polyester or water-intensive cotton. Research from Oxfam reveals how buying just one secondhand pair of jeans and a T-shirt rather than buying new could help to save the equivalent of 20,000 standard bottles of water, due to the resource-intensive production and processing associated with cotton.

Shifting attitudes on reuse

Studio ThusThat (pp. 14 and 36) has boldly predicted that no designers will be using synthetic, virgin materials in 50 years' time—a shift requiring widescale systemic change and transitions in cultural values. The case studies provide an insight into what this future could look like, a radical departure from how we now conceptualize and produce objects, systems, and architecture. This will require a material-first approach, flipping the traditional design process on its head. Typically, designers first consider the form and shape of an object or building, only to later explore which materials can be applied based on technical, functional, and aesthetic requirements. By reframing this approach, available and local resources drive the design process, and a material's properties are assessed upfront. The projects illustrate the ways that designers are already reimagining this future. Projects like Border (p. 194), which reuses workshop offcuts, and Dailly Courtyard House by Mamout Architects (p. 202), which repurposes on-site building components, demonstrate how materials can drive design innovation.

A patchwork wall of reclaimed bricks at Resource Rows by Lendager. The project addresses how we can recycle bricks with modern cement-based mortar.

Border is a minimalist furniture collection designed by Rikiya Toyoshima and Shomu Taki. The pieces use polycarbonate offcuts from a workshop in Okayama, Japan.

“By reusing materials that have lost their perceived value, we slow resource loops and extend their useful lives.”

Top: A lamp crafted by Lucas Muñoz reuses over 100 fluorescent tubes salvaged from old office ceiling lighting. Above: A door handle also by Lucas Muñoz.

Top: The Dailly Courtyard House by Mamout Architects uses a palette of on-site reclaimed materials. Above: The Circular Pavilion by Encore Heureux in Paris makes use of salvaged resources.

Integrating circular principles of long life, loose fit

Soundbounce is an acoustic installation by Mathilde Wittock that creatively tackles the environmental impact of discarded tennis balls, driven by her own personal sensitivity to noise.

Project
Soundbounce

Designer
Mathilde Wittock

Material
Used tennis balls

Resource Type
Sporting goods waste

Country of Origin
Belgium

Manufacturer
Mathilde Wittock

Area of Use
Furniture design

Category
Reused

The project seeks to address, in part, waste generated by the approximately 400 million tennis balls discarded each year, of which only 1% are reported to be recycled. Despite requiring up to five days and around 24 steps to produce, a tennis ball typically lasts just nine games. Current recycling efforts often involve the complete transformation of the balls, burning the felt and grinding the rubber, which compromises the materials' technical properties.

Wittock offers a more sustainable solution by collecting used tennis balls from local sports clubs in Belgium and repurposing them into sensory design elements such as wall paneling, furniture, and acoustic partitions. Each square meter incorporates approximately 283 tennis balls, the equivalent of 18¾ lb (8.5 kg) of carbon dioxide.

After the balls are cut along their white lines and dyed, they are placed into a wooden assembly with predrilled holes. The repeated pattern of folded balls creates a rolling surface that acts as a circular alternative to conventional textiles, also offering excellent sound absorption across a broad frequency range.

Through careful hand-cutting, dyeing, and assembly, Soundbounce transforms waste into visually engaging functional furniture pieces, including acoustic room partitions, wall frames, lounge chairs, and benches.

Addressing waste-management issues through upcycling

Local craftsmen in Zanzibar have transformed glass discarded by tourists into hotel furniture and souvenirs. Visitors can purchase these objects, connecting the origin of the problem to the solution.

Project
bottle up

Designers
Super Local, Klaas Kuiken, OSΔOOS, and Front Materials

Material
Glass bottles

Resource Type
Hospitality waste

Country of Origin
Tanzania

Manufacturer
Foundation bottle up Zanzibar

Area of Use
Furniture and product design

Category
Reused/recyled

Zanzibar is a picturesque Tanzanian island that attracts visitors from around the world with its rich culture and beautiful scenery. Large quantities of alcohol are imported each year, specifically for tourists. The island is unequipped to manage the quantity of waste glass left behind, however, and as a result these bottles often end up littering the landscape or mounting up in landfill.

Design studios Super Local, Klaas Kuiken, OSΔOOS, and Front Materials worked directly with local residents to address the problems posed by glass on the island. The team established a small factory where the discarded glass from local hotels could be repurposed. In just one year, enough glass was collected to fill six shipping containers. Unlike plastic, glass is fully and infinitely recyclable, making it a prime candidate for reuse. Recycling glass offers significant carbon savings, with a single ton saving up to 660 lb (300 kg) of carbon dioxide emissions.

The empty bottles collected were transformed into a collection of upcycled products, which integrated local crafts such as woodwork and weaving to further mask the glass's origin. The collection features light fittings, candleholders, and drinking glasses for tourists to buy—so they can take their waste home with them.

"The empty bottles collected were transformed into a collection of upcycled products, which integrated local crafts such as woodwork and weaving to further mask the glass's origin."

The design team devised ways to use every part of the bottle; even the dirty pieces and broken shards could be crushed and mixed with cement and water to create a terrazzo-like surface. At the time, most hotels on the island had been importing their furniture from far afield, so the team began designing locally made furniture from the glass such as benches and tables that local hotels could purchase. In addition, the design team discovered that Zanzibar imports construction sand from Australia because local sand is too soft to use. The team again crushed more broken glass, this time to a sand-like consistency, and mixed it with cement to make bricks. In 2022, the local team built their first house from the bricks.

Designing for circularity with a material-first approach

Border is a minimalist furniture collection made with the polycarbonate offcuts from a workshop in Okayama, Japan.

Project
Border

Designers
Rikiya Toyoshima and Shomu Taki

Material
Polycarbonate

Resource Type
Construction waste

Country of Origin
Japan

Manufacturer
Tanakakaken Co.

Area of Use
Furniture design

Category
Reused

The somewhat unconventional design process follows a material-first approach that ensures that form is driven by resource availability, instead of starting with form and considering materiality second.

The workshop carpenters had been using twin-wall polycarbonate, a sheet material made from two thin plastic layers with a hollow, ribbed core. It offers good thermal insulation and rigidity, and it is conventionally used for doors and windows in place of glass. Over time, the workshop accumulated sheets of various sizes due to the different dimensions of each custom-made door and window.

Designers Rikiya Toyoshima and Shomu Taki were able to rescue the offcuts from going to landfill by standardizing the stockpiled sheets into a set of uniform sizes before exploring circular design solutions. They developed a minimalist furniture collection working with the available resources, featuring stools, screens, lights, and tables. Caulk, which is typically used to seal air gaps around windows, was used to join the boards at their ends, emphasizing the rhythm and vertical lines of the polycarbonate.

A kit of parts made from modified everyday objects

Casa Umbrella is a building system for use in forming a temporary shelter, inspired by objects found in a common shop or market.

Project
Casa Umbrella

Designer
Kengo Kuma and Associates

Material
Umbrellas

Resource Type
Household items

Country of Origin
Italy

Manufacturer
Self-build

Area of Use
Architectural design

Category
Reduced

After exploring the suitability of various household items for use in architectural design, Kengo Kuma and Associates selected the humble umbrella as a primary building element. Among its unique qualities, the umbrella is easy to carry, lightweight, and importantly, water-repellent.

The team developed a modified umbrella, with three extra triangular segments, that could serve both in daily life and in a disaster kit. In the event of an earthquake, 15 people could come together with 15 of the umbrellas to build a covered space, in a social process requiring little skill or time, and no specialist tools.

The spherical structure, which references Buckminster Fuller's geodesic domes, is made up of triangular facets working in tension, supported by the skeletal ribs of the umbrella. Each umbrella is connected by waterproof zips, typically used for diving suits. These zips are sewn directly into the edges of the umbrella's surface, a highly waterproof, nonwoven, polyethylene fabric called Tyvek, produced by DuPont.

Project
Tide

Designer
Stuart Haygarth

Material
Plastic

Resource Type
Mixed ocean plastic

Country of Origin
United Kingdom

Manufacturer
Stuart Haygarth

Area of Use
Lighting design

Category
Reused

A chandelier elevating the discarded object

The first edition of the Tide chandelier was made from pieces of translucent ocean plastic washed up onshore. The spherical shape references the moon, which affects the tides that wash this debris onto the coastline.

Every year, thousands of people in the United Kingdom flock to the coast for the Great British Beach Clean, organized by the Marine Conservation Society. In 2023, the volunteers cleared a staggering 385 pieces of litter for every 330 ft (100 m) of beach.

Stuart Haygarth has been working with this very debris since 2005. His first piece, the Tide chandelier, was painstakingly crafted from discarded plastic washed up along Dungeness Beach in southeast England. Pieces of varying shapes and sizes are assembled to produce one perfect sphere, juxtaposing order and symmetry with the random nature of found objects. The end result elevates the sum of its parts to an object of value and beauty.

The work draws attention to the issues that plastic causes for our ecosystems and sea life. The U. K. government has estimated that plastic in our oceans kills 100,000 marine mammals and turtles every year, and the World Economic Forum predicts that marine plastics could outweigh all ocean fish by 2050.

Project
Dailly Courtyard House

Designer
Mamout Architects

Material
An existing warehouse

Resource Type
Construction waste

Country of Origin
Belgium

Manufacturer
RB Enterprises

Area of Use
Architectural design

Category
Reused

Viewing buildings as material banks for new construction

Reusing materials from existing buildings, or "urban mining," can significantly cut down on demolition waste, reduce the demand for virgin materials, and lower greenhouse gas emissions.

The reuse of building materials comes with numerous challenges, a major hurdle being the disconnect between supply and demand. Projects often operate under tight schedules and cannot rely on the availability of reclaimed stock; the somewhat unpredictable supply reduces the demand for large-scale reuse. Mamout Architects argues that architects must be opportunistic in what is often a volatile and uncertain reuse market, saying that "if you see a material or piece of equipment you like, buy it, and you'll find a way to incorporate it into the project."

For their Dailly Courtyard House project, the architects salvaged materials from a disassembled warehouse that once occupied the site, reusing bricks, steel beams, and glazed tiles. The industrial materials are thoughtfully chosen. However, on the selection of materials, the studio explains that you must let yourself be guided by what you find, advising that "there's no such thing as 'bad taste' in reuse, only unusual combinations."

Situated between two courtyards, Dailly's minimalist exterior is punctuated by expansive windows, revealing areas of exposed, reclaimed brickwork and glazed tiles.

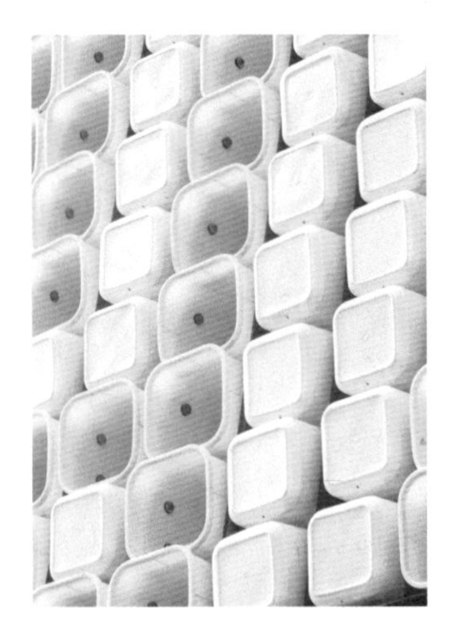

Adapting ice cream tubs to create a rainscreen cladding

More than 2,000 used ice cream tubs became a cost-effective, lightweight, and durable solution to cladding the Bima Microlibrary in Indonesia.

Project
Bima Microlibrary

Designer
SHAU

Material
Ice cream tubs

Resource Type
Household waste

Country of Origin
Indonesia

Manufacturers
Yogi Pribadi and Pramesti Sudjati

Area of Use
Architectural design

Category
Reused

As part of efforts to combat high illiteracy rates in Indonesia, the 1,700-sq-ft (160-m^2) Microlibrary was conceived as a prototype for a series of small libraries across the country, made with locally sourced materials.

SHAU first considered jerry cans as a cladding material, but unable to source enough, found an online vendor listing used ice cream tubs. These tubs were mounted to a steel frame and tilted downward, to divert rainwater and protect the building from weathering. During tropical storms, sliding-glass doors behind the facade can be closed, creating a fully weatherproof barrier.

Some of the tubs have had the bottoms cut out, leaving them open to allow natural ventilation and light to filter into the building. The architects used a variation of closed and open tubs to embed a binary message that can be read from the outside. As it would have been time-consuming to adapt each plastic container, the architects engaged local craftsmen, who were able to produce their own tools, with which they used to cut out the bottoms of tubs in a quick and simple manner that also provided a high-quality finish.

The building was placed above a modest, elevated platform, an existing community feature. The gesture enhanced the locale's public value, with the gathering place sheltering below the first-floor library.

Rejecting superfluous details and surplus materials

Located in a former theater and recording studio, this restaurant was conceived to reevaluate patterns of consumption, minimize use of virgin materials, and avoid superfluous actions that are aesthetically driven.

Project
Mo de Movimiento

Designer
Lucas Muñoz

Material
Finishes and fittings

Resource Type
Construction waste

Country of Origin
Spain

Manufacturer
Zimenta Construcciones

Area of Use
Interior design

Category
Reused

After opening up the space and creating an internal courtyard, up to 1.9 tons (1.7 tonnes) of construction rubble were used to make seating and backrest modules for rows of benching. Building debris was turned into large-format terrazzo blocks using casting molds on-site—each block is left unfinished, except for one strategically polished surface. The chairs and tables throughout the space are made from the reclaimed pinewood boards of the building's former auditorium seating.

Every intervention serves a practical purpose, including two large woodburning terracotta ovens, which form the heart of the building's heating system. The system is supported by a network of copper conduits that help recover the calorific energy from the fire (which would otherwise dissipate) and distribute it throughout heated floors and radiators. The building also uses a low-impact cooling system based on the natural adiabatic exchange of heat between hot, dry air and wet environments. Nine suspended terracotta jugs handcrafted by artisan Antonio Moreno Arias have been shown to lower the local temperature by up to 59 °F (15 °C).

EXTINTOR
SALIDA

Social value is embedded in the project, which relies on local initiatives and suppliers for its material streams, collaborators, and food for the restaurant itself.

A number of the restaurant's lights are rescued from car parks, their fluorescent bulb cases brought up-to-date with LED technology by students from the Norte Joven Association and the design team.

A sustainable restaurant curating waste from the city

Located in Bangalore, Big Top, better known as the Circus Canteen, is a collage of discarded objects, exemplifying a grass-roots, democratic approach to design.

Project
The Circus Canteen

Designer
Multitude of Sins

Material
Various donated objects

Resource Type
Urban waste

Country of Origin
India

Manufacturer
Multitude of Sins

Area of Use
Interior design

Category
Reused

Multitude of Sins estimates that the Circus Canteen interior is made up of about 90% reclaimed materials, coming from multiple sources, including a nearby commercial building that was being refurbished and a citywide appeal for unwanted household objects.

The Circus Canteen embodies the Indian philosophy of *Jugaad.* A resourceful approach to problem-solving deeply ingrained in Indian culture, it rejects consumerism and makes do with what is available.

Visitors to the 2,134-sq-ft (198-m^2) restaurant are guided through a series of scrap metal archways, lit overhead by old automobile headlights suspended from bicycle chains; the flooring below is a patch-work assemblage of tile samples sourced from local ceramic showrooms. The walls of the restaurant are decorated with various signs and defunct electronics, and its secondhand furniture has been upcycled with maximalist motifs referencing the city's vibrant street culture.

MECHANIMAL
SHOW ME THE
HARMONY

Hot Meals
MOTEL

The Circus Canteen is part of Bangalore Creative Circus, a collective of artists, scientists, and change makers hosting community-focused events in the city.

Using an unconventional design process, Multitude of Sins crafted surface finishes, lighting, furniture, and art installations with materials from citywide donations, salvage markets, and dumping yards.

Mystical Brews
Fine Potions
No Stock

A microbrewery fostering community pride

Kamikatz Public House in Kamikatsu, Japan, combining a microbrewery, pub, and secondhand store, is a pioneering force in the town's zero-waste initiative.

Project
Kamikatz Public House

Designer
Hiroshi Nakamura & NAP

Material
Windows

Resource Type
Construction waste

Country of Origin
Japan

Manufacturer
Daiso

Area of Use
Architectural design

Category
Reused

The building organizes its functions in a linear order, visualizing the cycle of production and consumption: starting with rooms used for storing raw materials, moving to the boiler chamber and brewery, and ending at the pub and store, where goods can be bought. Rather than concealing the production process, the building's design speaks to the fact that ignorance around the labor and resources required to create a product contributes significantly to our growing waste problem.

Architecturally, windows have been salvaged from abandoned houses in the town to form an 26-ft- (8-m-) high patchwork facade in the bar area. Carefully removing and transporting these easily shattered glass components requires effort—something most people overlook. But the architects recognize the windows' symbolic importance, representing a connection between the town and the structure. Functionally, the windows help to reduce heat loss in the winter, trapping air between the facade's dual layers. In the summer, warm air is passively drawn up the double-height space and can be expelled at the top.

RISE & WIN
BREWING CO.
BBQ & GENERAL STORE

The Kuru Kuru shop at Kamikatsu's waste collection center displays used items for locals to take home at no cost; this enables the reuse of approximately 1,200 lb (550 kg) of discarded items per month.

Project
Stüssy Boucherouite T-Shirt Rugs

Designer
Artisan Project

Material
Dead-stock T-shirts

Resource Type
Textile waste

Country of Origin
Morocco

Manufacturer
Ain Leuh Women's Cooperative

Area of Use
Soft furnishings

Category
Reused

Handwoven rugs made from old T-shirts and local wool

For its second collection of boucherouite rugs, Stüssy has collaborated with Artisan Project, a collective headed by Palestinian-American textile designer Nina Mohammad.

Boucherouite is a Moroccan-Arabic term meaning a piece of torn and reused textile. This collection exemplifies the rug-making tradition of North Africa, where Berber women have handwoven boucherouite rugs for generations. The project's rugs are composed of upcycled Stüssy T-shirts and wool from the Atlas Mountains.

When T-shirts are ripped, marked, or stained during the manufacturing process, clothing brands like Stüssy accumulate large volumes of defective clothing. This project aims to resolve this stagnancy, recycling T-shirts with manufacturing irregularities to stimulate an ecosystem of dead-stock fabric. During its first partnership with Artisan Project, Stüssy was able to repurpose about 650 lb (300 kg) of old T-shirts.

Nina Mohammad draws inspiration from the natural beauty of Morocco's landscapes and diverse flora as well as the region's traditional rugs, art, and visual design. The rugs are handwoven by members of the Ain Leuh Women's Cooperative in Morocco, using deeply embedded cultural practices. The women use a vertical loom to weave the textiles, with only a pair of clippers and a comb to keep fabrics from tangling.

The designs in the collection take cues from the picturesque landscapes of Morocco and the area's storied rug-making heritage, honoring traditional craftsmanship and artisan practices.

A pavilon in Paris reconsidering supply chains

Built to coincide with the opening of the 2015 United Nations Climate Change Conference, held in Paris, the pavilion draws attention to the possibilities of reusing building components.

Project
The Circular Pavilion

Designer
Encore Heureux Architects

Material
Doors

Resource Type
Construction waste

Country of Origin
France

Manufacturer
Cruard Charpente

Area of Use
Architectural design

Category
Reused

Back in 2006, the reuse consultancy organisation Salvo estimated that each year approximately 2 million doors were disposed of in the United Kingdom alone; this number has decreased, but huge potential remains for reuse.

Just under 10 years on, in May 2015, three months prior to the Circular Pavilion's completion, Encore Heureux learned of an opportunity to salvage 180 oak doors from a housing refurbishment project in Paris. Upon seeing the doors, the architects immediately changed course, allowing the found materials to lead the design process. Arranged in a herringbone pattern, the doors became one of the main drivers of the structure and its jagged, saw-tooth roof form.

Reclaimed materials make up about 80% of the building. Its mineral-wool insulation was recovered from the roof of a supermarket, and the timber structural framing was rescued from the construction site of a retirement home. Within the building, the floors and interior partitions were formerly exhibition walls, while the 50 wooden chairs were reclaimed from local recycling centers.

Preserving the historical elements of a rural family home

Gonzalez Haase AAS has converted an old German farmhouse for contemporary living, with concrete, wooden roof trusses, and oriented strand board (OSB) insertions.

Project
The Four-Window House

Designer
Gonzalez Haase AAS

Material
Concrete fragments, wood, and OSB

Resource Type
Existing building fabric

Country of Origin
Germany

Area of Use
Architectural design

Category
Reused

Nestled into the rural landscape, three farm buildings are arranged around a U-shaped courtyard. Two of the structures were already restored; this project focuses on the third—the family home. The original house underwent low-cost alterations in the 1980s, leaving it with limited preservation value.

Due to its poor condition, the architects carefully stripped away the redundant layers, retaining only the historical elements of the original structure. These pieces were thoughtfully incorporated into a new, cohesive concrete structure, with the scars of this intervention left exposed and legible.

The roof soffit, interior partitions, and formwork of the concrete walls were all shaped using OSB, imparting a textural imprint to the concrete surfaces. Although concrete is carbon-intensive, the architects used it sparingly, prioritizing durability and longevity. OSB utilizes entire trees, including irregular, deformed, and knotty wood strands that would otherwise be discarded. The house adopts a minimalist approach, with no internal doors and all surfaces left untreated—without paints, varnishes, or paneling.

Each facade features a large opening that functions as both a window and a door, thoughtfully positioned to maximize natural light around the floor plate.

Exploring buildings as opportunities for urban mining

Resource Rows challenges the construction industry's linear model of consumption, reclaiming building materials from local projects at end of life, despite challenges to the reuse of modern materials.

Project
Resource Rows

Designer
Lendager

Material
Upcycled bricks, waste wood, a recycled concrete beam, and old windows

Resource Type
Construction waste

Country of Origin
Denmark

Manufacturers
AG Gruppen

Area of Use
Architectural design

Category
Reused

The project repurposes waste timber, old windows, and even uses a former concrete beam as a bridge between two rooftops, reducing resource use and the amount of material sent to landfills. The reuse of bricks from abandoned buildings, however, came with particular challenges due to the use of modern construction materials.

Since the 1960s, it has become harder to salvage bricks. Around this time, lime mortar was phased out in favor of stronger, cement-based mortar that created stronger bonds and also made it difficult to recover individual bricks without cracking them. Lendager used angle grinders to cut the bricks out in 10.8-ft (1-m^2) panels that could be mounted into a steel frame and integrated into a new patchwork facade—saving 1.1 lb (0.5 kg) of carbon dioxide per rescued brick.

Studies indicate that Resource Rows has prevented 510 tons (463 tonnes) of waste from going to landfills, without adding cost. While the circular economy is still emerging, Lendager argues that sustainable solutions will need to be cost-neutral, or better, beneficial, for developers to adopt them.

STORSKRALD

Although sustainability is still broadly assumed to add cost, Lendager saved 510 tons (463 tonnes) of waste materials without increasing the construction budget.

Reframing off-the-shelf materials in a workspace

Lucas Muñoz worked collaboratively to transform a workspace on the fourth floor of Madrid's iconic O'Donnell building into a creative laboratory for product design firm Sancal.

Project
Sancal

Designer
Lucas Muñoz

Material
Finishes and fittings

Resource Type
Construction waste

Country of Origin
Spain

Manufacturers
Zimenta Construcciones and Zetus

Area of Use
Interior design

Category
Reused

Items found in the fourth-floor space that could not be conventionally recycled were prioritized for reuse, as they would otherwise go to landfill. This made for a radically logical approach that required a highly creative design process.

For example, the designer found that the old, raised-access flooring system was composed of plastic veneer, a chipboard core, and a thin backing layer of aluminum, none of which could be easily separated and recycled. Instead, the aluminum backing was buffed and transposed to produce a metallic matte finish to two walls and a new ceiling margin. About 270 of the original 350 flooring cassettes were preserved; only those with significant damage or irregular geometries were discarded. A number of the pedestals from the flooring system were used as coat-racks and cupboard handles, however, because they are, iron many could be easily recycled.

The suspended ceiling tiles of the original interior were 100% nonrecyclable, and therefore prioritized. The recovered modules were snapped in half, leaving two tiles, each with a jagged edge, that could be repeatedly stacked and covered in plaster, creating a feature textured wall finish.

“Items found in the fourth-floor space that could not be conventionally recycled were prioritized for reuse.”

About 40 fluorescent light boxes were recovered and rewired with LED technology by students from the Asociación Norte Joven, an organization that offers training to young people at risk of social exclusion. These boxes were repurposed into modular luminaires that hang from adjustable hooks. The fluorescent tubes themselves are a challenge to recycle because of their fragile and often contaminated glass and aluminum components, which contain toxic chemicals such as mercury. For this reason, the design team wanted to extend the tubes' useful lives, choosing to use them as screens to diffuse the new LED light source.

The original finishes within the central enclosed meeting room have been preserved, serving as a tangible reminder of the space's history. This gesture offers a cross-section of time, presenting past architectural languages and revealing the origins of the circular interventions carried out by Muñoz and his team.

To define appropriate technical solutions and strategies, Muñoz worked closely with a team of experts including Tomás Miranda from Zimenta Construcciones as well as Joan Vellvé, Wilson and José Luis, and Rafael Abad.

Project
Casa Quinchuyaku

Designer
Emilio López

Material
Reclaimed building materials

Resource Type
Construction waste

Country of Origin
Ecuador

Manufacturer
Self-built

Area of Use
Architectural design

Category
Reused

A community-built house that promotes responsible living

Part of a larger project to gradually replace eucalyptus plantations with native flora and fauna, Casa Quinchuyaku is located 9 miles (15 km) from Quito, on the eastern slope of Ilaló, an extinct volcano 8,500 ft (2,600 m) above sea level.

Since the native vegetation was eradicated to make way for eucalyptus tree plantations, forest fires have intensified, particularly during the region's annual dry season. The slopes of Ilaló have also been increasingly affected by soil erosion in recent years. Casa Quinchuyaku seeks to restore biodiversity and enhance soil quality through sustainable practices—using recycled greywater from bathrooms and kitchens for irrigation, and converting human waste into compost for planting. Additionally, a system of terraced landscapes with infiltration trenches mitigates erosion by slowing rainwater runoff, allowing water to gradually permeate the soil rather than washing it away.

The fully solar-powered house was constructed through a series of minga gatherings, an Inca tradition of community labor, fostering an environment of learning and sharing on-site. Approximately half of the construction materials used are reclaimed components: the stairs, countertops, beams, columns, and every door and window originate from a former residence in the heart of Quito.

The architecture is conceived as part of the landscape and its geography. A double-opening envelope works as a mediator in the landscape and allows for cross ventilation.

Index

Lucas Muñoz
lucasmunoz.com
Mo de Movimiento
Photography:
Lucas Muñoz Muñoz
pp. 212–217

Sancal
Photography:
Lucas Muñoz Muñoz
pp. 187 left, 244–247

N

Hiroshi Nakamura & NAP
nakam.info
Kamikatz Public House
Photography:
courtesy of Nacása & Partners Inc./
nacasa.co.jp
pp. 224–227

New Makers Bureau and Localworks
newmakersbureau.com
localworks.ug
32° East Arts Centre
Design team:
James Hampton, Laura Keay,
New Makers Bureau
Project architect:
James Hampton
Local architect, structural engineer,
cost consultant, M&E engineer:
Localworks
Photography:
Tim Latim/@timothy_latim
pp. 113, 114 bottom, 115, 116, 117 top left
Will Boase/willboase.com
pp. 112, 114 top, 117 top right, bottom

newtab-22
newtab-22.com
Sea Stone
Photography:
courtesy of newtab-22
(Hyein Choi, Jihee Moon)
pp. 120, 136–139

P

Harry Peck
@harrypeckstudio
Wave Cycle
Photography:
courtesy of Harry Peck Studio
pp. 11, 46–47

Jorge Penadés
penades.xyz
Structural Skin
Photography:
Brenda Germade
pp. 34, 35 bottom
Gonzalo Machado
p. 35 top

S

Shakúff
shakuff.com
Crystal Shell Collection
Photography:
courtesy of Shakúff Bespoke
pp. 56–57

Shau
shau.nl
Bima Microlibrary
Photography:
Sandy Pirouzi Sanrok Studio/
sanrokstudio.com
pp. 208–211

James Shaw
jamesmichaelshaw.co.uk
Plastic Baroque
Photography:
Paul Plews/paulplews.com
pp. 6, 42–45
Rory Mulvey/rorymulvey.com
p. 12 left

Snøhetta
snohetta.com
ateljelyktan.se
Superdupertube
Photography:
courtesy of ateljé Lyktan
pp. 180–183

Studio Flaer and the National Taiwan Craft Research and Development Institute
studioflaer.com
ntcri.gov.tw
Indigo
Photography:
courtesy of Studio Flaer GmbH
pp. 168–171

Studio Johanna Seelemann
johannaseelemann.com
Potentials
Photography:
courtesy of Studio Johanna Seelemann
pp. 38–39

Studio Kloak
studiokloak.com
The Blueprint Collection
Photography:
courtesy of CoCo Ree Lemery
and Studio Kloak
pp. 160–161

Studio ThusThat
thusthat.com
Red Mud
Photography:
courtesy of Studio ThusThat
pp. 14–15

This Is Copper
Photography:
courtesy of Studio ThusThat
pp. 10, 36–37

Super Local, Klaas Kuiken, OSΔOOS, and Front Materials
super-local.com
klaaskuiken.nl
osandoos.com
front-materials.com
bottle up
Photography:
Jeroen van der Wielen/
jeroenvanderwielen.nl
pp. 192–195

T

The New Raw
thenewraw.org
Knotty and Pots Plus
Photography:
Stefanos Tsakiris/stefanostsakiris.com
pp. 13 left, 17, 19 bottom
Mathijs Labadie/
mathijslabadie.com
pp. 16, 19 top left
Federico Floriani/
federicofloriani.com
p. 18
Micheèle Margot/
michele-margot.com
p. 19 top right

Circular Materials

Innovation and Reuse in Design and Architecture

This book was conceived, edited, and designed by gestalten.

Edited by Robert Klanten and François-Luc Giraldeau
Contributing Editor: Joe Gibbs

Text by Joe Gibbs

Editorial Management: Arndt Jasper
Editorial Assistance: Effie Efthymiadi

Design, layout, and cover: Stefan Morgner
Photo Editor: Madeline Dudley-Yates

Typefaces: Trade Gothic Next by Jackson Burke, Akira Kobayashi, and Tom Grace

Front cover image:
Séverin Malaud / severinmalaud.com / @severinmalaud

Back cover images:
Top left Magnus Dennis / magnusdennis.com;
center left Johanna Hehemeyer-Cürten and Lobke Beckfeld / @johlemo / lobkebeckfeld.com, courtesy of Fernando Laposse / fernandolaposse.com; bottom left Thomas Meyer / Ostkreuz Photography / thomas-meyer.com;
right courtesy of VANK / vank.design

Printed by Grafisches Centrum Cuno GmbH & Co. KG, Calbe (Saale)
Made in Germany

Published by gestalten, Berlin 2025
ISBN 978-3-96704-175-0

1st printing, 2025

For more information, and to order books, please visit www.gestalten.com

Die Gestalten Verlag GmbH & Co. KG
Mariannenstrasse 9–10
10999 Berlin, Germany
hello@gestalten.com

Bibliographic information published by the Deutsche Nationalbibliothek. The Deutsche Nationalbibliothek lists this publication in the Deutsche Nationalbibliografie; detailed bibliographic data is available online at www.dnb.de

None of the content in this book was published in exchange for payment by commercial parties or designers; gestalten selected all included work based solely on its artistic merit.

This book was printed on recycled paper certified according to the standards of the FSC®.